D1026628

Christopher and Dolores Lynn Nyerges

# extreme simplicity

A Guide to
Urban Homesteading

Dover Publications, Inc.
Mineola, New York

## Copyright

Foreword Copyright © 2013 by Christopher Nyerges
Copyright © 2002 by Christopher and Dolores Lynn Nyerges
Unless otherwise noted, illustrations copyright Barbara Carter
All rights reserved.

## Bibliographical Note

This Dover edition, first published in 2013, is an unabridged republication of the work originally published by Chelsea Green Publishing Company, White River Junction, Vermont, in 2002 under the title and subtitle *Extreme Simplicity: Homesteading in the City*. A new sixteen-page color insert, located between pp. 208 and 209, has been added to the present edition. All photographs are courtesy of the authors. A new Foreword to the 2013 edition has been specially prepared for this volume by Christopher Nyerges.

## Library of Congress Cataloging-in-Publication Data

Nyerges, Christopher.
    Extreme simplicity : a guide to urban homesteading / Christopher and Dolores Lynn Nyerges. — [Rev. ed.]
        p. cm.
    Guide to urban homesteading.
    "This Dover edition, first published in 2013, is an unabridged republication of the work originally published by Chelsea Green Publishing Company, White River Junction, Vermont, in 2002 under the title and subtitle Extreme Simplicity: Homesteading in the City."
    Includes bibliographical references and index.
    ISBN-13: 978-0-486-49114-1
    ISBN-10: 0-486-49114-5
    1. Urban homesteading—California—Los Angeles. I. Nyerges, Dolores Lynn, 1946– II. Title. III. Title: Guide to urban homesteading.

HD7289.42.U62L676 2013
640—dc23

2013010226

Manufactured in the United States by Courier Corporation
49114501    2013
www.doverpublications.com

Live light upon the land
if you would not be earthbound.

—Shining Bear

# Contents

# Acknowledgments

We are very thankful to have had in our lives the influence of Richard E. White, founder of the Rainbow School and the nonprofit educational think-tank WTI. Through our classes and meetings and seminars with him, we have developed a certain mind-set of "living lightly on the land" that has become second nature to us. Many of the things we now do and many of the systems we now employ in our household were directly or indirectly inspired through our interactions with White. We are extremely grateful to have had that friendship and guidance.

Over the years, various other experts and programs have influenced our thinking. We name some of these individuals in the book, but there are far too many over too many years to mention individually. To all of you we extend a big "thank you"!

# Foreword to the 2013 Edition

It has been well over a decade since the first release of our *Extreme Simplicity* book, where we documented our nearly two decades of "living lightly in the city," in the hilly backwoods of Los Angeles.

Dolores and I wrote about how we lived, and what we did, to practice an ecological and economical lifestyle. Our blueprint for life was that everything we did, and all the resources we interacted with, should be karma-positive, or karma-neutral. By this I mean that we judged our actions by how self-reliant we became, and whether or not our actions were a part of the solution for the salvation of the Earth.

We used this guideline to consider whether or not to have a lawn, how to deal with trash, what to plant, how to deal with utilities, what sort of employment to engage in, etc. We were not wealthy, nor were we the smartest. So under those circumstances, this book is our testament to those decisions and actions we took.

A lot has changed since *Extreme Simplicity* was first released in 2002. We have witnessed eco-living go nearly mainstream, as folks in increasing numbers are replacing the front lawn with a garden. Voluntary simplicity, sustainable landscaping, organic gardening, and urban farming have all established solid roots around the country. We're glad to have been a part of that mindset.

In 2008, Dolores passed away, which was a big loss to me personally, and to all those who knew and loved Dolores. But she lives on in this book, which embodies so much of her words and thinking.

Despite clear signs of progress, there is still much fear and cynicism in the world. Do all you can to not let fear, and panic, and cynicism, control your thinking and your actions. Accept that you cannot change the world, but that you can at least work on changing yourself. Be a part of the solution. Think globally and act locally. And as Gandhi said, "Be the change that you wish to see in the world."

CHRISTOPHER NYERGES
*October 2012*

# Introduction

*I prize even the failures and disillusionments which are but steps towards success.*
—GANDHI

There are many books written by people who were sick of modern society, and who moved to the country, built their house, and started growing their own food. That's good—it's just not what we, Christopher and Dolores, have chosen to do.

Simply put, we have chosen to live lightly on the earth right here in the city, and to do so in a way that represents solutions to the problems that today confront everyone. Our way of living more lightly on the earth has been described by many names: voluntary simplicity, living country in the city, an ecology lifestyle, and so forth. We regard what we do as practical survival.

Both of us are from the San Gabriel Valley, the eastern side of Los Angeles County. We are both from the city and have many roots here.

But we have each lived in remote rural settings as well. Since early childhood, we have been drawn to faraway, less crowded places. In part, this was a result of growing up in the crowded suburbs and cities of Southern California. And we've learned many valuable lessons from the wilderness, from farming communities, from the Hawaiian backcountry, from the silent deserts and pure beaches—the places where we dreamed about living and being.

When our paths crossed in 1980, we were each in our own ways involved in sharing a more natural way of life through teaching about nutrition, wild foods, and home self-reliance. Christopher was interested in Native American skills and beliefs, backpacking, wild edibles.

Dolores was interested in intentional communities, food storage, gardening, spiritual seeking.

We both became affiliated with a nonprofit "think-tank" organization, WTI, Inc., whose fundamental premise is that in today's world there is nowhere to run away to. WTI's purpose was, and continues to be, to research and educate in all areas of "survival." Survival, of course, has many levels of meaning, and the goals and aspirations of this group have coincided with our own. The founder of WTI has been influential to both of us in the life choices we have made, and we continue to learn with and work with this unique organization (see Resources for contact information).

We have long been aware of the environmental degradation occurring everywhere, resulting in pollution, trash, toxic landfills, poisoned water, and disappearance of wilderness. Of course, the ongoing crisis of overpopulation is what drives these other environmental dangers, and certain forms of perverse thinking underlie all the separate problems that we see in the world.

In the 1960s and 1970s, the two of us found many avenues of "escape," such as travel, backpacking, living a peaceful existence in a remote area, and so on. But we have not been content merely to escape. Escapism is a controlled fantasy. You can never really escape the extant reality.

So then, what does one do?

We have come to realize that it is not possible to change the world. It is hard enough work to change one's own thinking and to actually live one's life in a manner that represents a solution to at least some of the world's problems. As an outgrowth of the teaching we had already begun, we founded the School of Self-Reliance (www.self-reliance.net). In our classes, seminars, and consultations, we have spent thousands of hours exploring what makes people tick and what radical action concerned and dedicated individuals can take, in both our thinking and actions, to align ourselves with the Plan of the Universe. This is not an easy task, and though our failures are many, this is what we have endeavored to do in our personal lives and via the many classes we teach.

After living in rural areas, we each moved back to the Los Angeles area. We found ourselves living here because we are able to work with and to reach others. We realized that this is perhaps where we could be of the most benefit. Neither of us felt that we had such opportunities in the more peaceful, remote, faraway places we had visited and lived.

It's important to emphasize that our thinking is not like those who feel there is nothing of value in "modern civilization" and that everyone should therefore flee from the cities. One only has to study the development of modern cities to see that the opportunities for coming together in groups are many in urban areas. For instance, in cities the arts and education can flourish, because you can bring together many of the best teachers. Cities are also places where trade and innovation can flourish, and they are places that provide quick communication and many kinds of protection.

But there are dark sides to consider, as most know only too well, when the numbers of people packed together increase beyond a certain level.

Our involvement with WTI and its various activities has helped us recognize that we are already at a crisis level in many areas. Those who would survive in strength need to do so with eyes fully open, proactively, not waiting or hoping for "the government" to solve problems.

One of the early WTI public activities was the Noah Seminars, named, of course after the Noah who built an ark to survive the Great Flood described in the Bible. Humankind's penchant for avoidance—and for "seeing what we want to see"—renders most of us incapable of actually seeing the crisis we're in. Participants in the Noah Seminars actively looked for solutions that could be implemented within their own family lives, solutions that were within the law and would be possible within the existing social and political systems. In other words, the solution was never to run away and form a commune or some other such action that would be viable only for a handful of people. If there were to be an ark for the equivalent of the modern-day "flood" that may destroy us all, the new ark would be our wisdom, our training, our skills, our thinking, our timing, our actions.

Could problems be identified and solutions be found that could be employed by anyone, anywhere? Participants of the Noah Seminars endeavored to do just that, and they identified specific areas for exploration. These included

1. internal physical health
2. physical strength and coordination
3. physical safety
4. mental health
5. societal health
6. environmental health
7. economic health
8. sexual health
9. urban health
10. ethical or moral health
11. spiritual health

These are the areas where the actual crises facing all of us, right now, take place. What the two of us do, day in and day out, is explore and study and research the nuances of each of these so we can gain clarity as to the most desirable actions for survival. Of course, this book does not attempt to cover the global crisis in that scope; we are sharing this as background, to provide a context for the more specific descriptions we give here of our home, gardens, animals, and self-reliance strategies.

It is always a shock to us to encounter people who confidently believe that their way of life in the United States will always be what it is today—that there will always be plenty of food and available electricity, that there will always be relatively safe neighborhoods, banks you can rely on, and hardware stores you can just drive to, to get whatever you need. We follow the news around the world, and it is sobering to see that there is always a group of people somewhere (often, many groups) whose entire way of life has been disrupted by war, by evacuations from floods or fires or earthquakes, or by political manipulation. Those who know the basic survival skills tend to fare the best. Despite all the technological achievements of our humanity, basic survival skills will never become irrelevant. None of our "higher aspirations" or "lofty goals" can ever be reached if we are unable to survive on a basic physical level.

Think about it: If all technological support systems were suddenly removed and you were alone in the Mojave Desert tomorrow, how much do you actually *know* about the way things work so that you could begin living your life above the "wild animal" level?

Keep in mind that a major earthquake or hurricane or epidemic—to say nothing of a nuclear exchange—would drastically compromise if not destroy the technological infrastructure of your comfortable 21st-century home.

So one has to begin from the beginning. Consider these questions:

1. Do you know how to locate potable liquids?
2. Do you know how to find wild foods?
3. Could you make shelter to protect you from 120-degree heat and 20-degree cold?
4. Could you deal with typical pains and discomforts without commercial medications? Do you know how to find wild medicines?
5. Could you avoid the consequences of overexposure to heat and cold? Dehydration? Hypothermia? Frostbite?
6. Could you keep your body clean, in terms of ordinary hygiene? Could you safely deal with daily normal eliminations? How about menses? Could you handle the strong body odors that would manifest over time?
7. How would you handle sunburn, windburn, or cracked lips?
8. Could you fabricate clothing (including hat and shoes)?
9. Could you make fire?
10. Could you fabricate cooking and eating utensils?
11. Could you make electricity?
12. Could you make a knife, ax, or bow and arrows?
13. Could you find a way to write and keep records?

Most urban people can answer yes to only one or two of the questions in this list, and almost never do we meet someone who can emphatically answer yes to all of them.

We can get even more basic. The three urban fundamentals of physical survival that are rarely addressed adequately when the topic of survival preparedness comes up are

- *the ready availability of food;*
- *the ready availability of medical aid, including doctors and hospitals; and*
- *waste-disposal systems.*

All of these basic urban services would be severely disrupted after an earthquake or other disaster. In living our own lives, we have put into practice many skills rarely practiced in cities that are now merely part of our daily routines. We know that these three areas—food, health, and waste disposal—are critically important, and many of the techniques we employ in our urban homestead address those necessities.

This book's purpose is to explain a variety of relatively simple, natural, and low-cost methods and techniques that we feel anyone can adopt, under most situations. We address all areas of the urban home, including the yard, sources of food and water, recycling, utilities and alternatives, animals in the city, and interacting with neighbors. We are sharing what we have done with our limited income and explaining why we do what we do.

Of course, we know that we can always improve upon our present solutions and that as we go along we will keep on learning. By no means do we live a static life here. We are constantly refining, researching, and observing, and we are quick to change something whenever we see that it no longer serves us well. We hope that anyone who is committed to becoming a part of the solution to the global crisis, in the process becoming increasingly more self-reliant and healthy, will benefit from the ideas presented here.

Feel free to contact us to discuss anything in this book. We have included our contact information in the resources section.

# 1  Our House

*Everybody thinks of changing humanity, and nobody thinks of changing himself.*

—Leo Tolstoy

When we purchased our home in the mid-1980s, it was one of the most dilapidated places in the neighborhood. Clearly we had work ahead of us. A duplex rental with a distant owner and careless tenants, the building had been sorely neglected. Yet we were glad to discover as those first weeks and months went by that the damages due to neglect were mostly cosmetic or easily repaired.

There were no serious problems with the building's structure, apart from a leaky roof, which we replaced as soon as we could afford to. The water pressure could be better, which would mean replacing some of the pipes. And the electrical system, though surely *fine* for the 1950s, would need to be modernized.

But our need to make these improvements gave us an opportunity to reconsider our priorities for the house in light of our longer-term goals. We wanted our various projects to steer us in the direction of self-reliance—even in an urban setting—and in the direction of living our lives lightly. All this we proposed to do within a modest budget on a city lot in Los Angeles.

The interior of the building was generally shabby, and the back section especially was very run-down. We retiled the front kitchen and bathroom and painted all the walls in the front section of the duplex so that we could rent it.

We removed the garbage disposals from both kitchens and put them in the city's recycling bin. These costly and noisy appliances aren't necessary and cause endless plumbing problems. Plus, think about it: What are we doing when we use a garbage disposal? Using extra water and extra electricity to grind up "garbage" so it can pass down the sewer lines and end up eventually in the ocean. Our choice is to give our food scraps to our animals.

We also removed the automatic dishwashers from each kitchen. We've heard interesting debates about whether these modern devices use more or less water than simply washing dishes by hand. Usually electrical use is not factored into such debates, and besides, we find that the quiet times spent washing dishes, looking out the window toward the chicken coop, is a waking meditation. We salvaged whatever hardware we could from the dishwashers and sent the rest to the recycling center.

Although the house came with two natural-gas wall heaters, one was dangerously corroded and we had it disabled. The other we have used only when necessary. Eventually we installed a fireplace in the back, unheated section. Many people have been surprised, even shocked, to learn that we do not have "modern" heating, but our merino wool sweaters and our fireplace are usually adequate during the Southern California winters.

Nor do we have "central cooling." In the summer, however, we discovered that this location has very still air, and we didn't get much of a breeze through the house. Partly this was due to the fact that we closed and locked our doors at night. Over the years, we replaced regular screen doors with steel security screens, so as to be able to leave the doors open to the air all night without worrying about a break-in. This has made a terrific difference, allowing cooler air to flow through the house.

We mentioned the old roof, which leaked terribly during winter rains, but we couldn't afford the expense of a new roof right away, though we were convinced that the existing dark brown roofing would keep the house much hotter in summer than it had to be. We researched the many "liquid rubber" products on the market. For a few hundred dollars, we painted the whole roof with a coat of white Roofer's Best, which is sold primarily for use on trailers to help keep down solar-heat absorption, and not as a roof sealant, though it did seal most (not all) of our leaks. Its main value has been to keep the house 15 to 20 degrees cooler during the summer than it had been with the dark roof. It is amazing to be inside a cool house, with no air conditioner, when outside summer temperatures are over 100 degrees Fahrenheit. Again, we are simply using natural principles—in this case, the reflective properties of a white roof. Eventually we had our roof professionally reshingled in the lightest color available.

Keep in mind that we live within the boundaries of a major metropolitan area. We are not living out in the country, nor are we living off-the-grid,

DOLORES LYNN NYERGES

*Ramah (left) howls while Cassie relaxes in front of the living room security door.*

supplying all our household power with solar, or going without electricity altogether. Yet we feel that all too many city dwellers have used an urban home as their excuse *not* to adopt some of the methods practiced by country people, thereby missing all kinds of opportunities for special learning and savings. Even given the constraints of urban life, we have tried to grow as much of our food and provide as much of our own oxygen as possible, recycling whatever we can, collecting rainwater, and living our lives with no excessive use of resources.

While the house wasn't in dire shape, the yard was a wasteland. The previous residents had used the front yard alongside the driveway to park and work on their cars, resulting in soil as hard-packed as adobe. Some weeds and crabgrass had managed to pop up here and there after rains, but otherwise it was barren.

The garage door was broken and falling off and had been propped up with a two-by-four. One of our first projects was to have an aluminum garage door installed, which is lighter, easier to operate, and needs no maintenance. Instantly we had another usable enclosed space.

3

The front courtyard was a hodgepodge of introduced ornamental plants. A stand of banana trees in a planter had never been thinned, and the expanding roots were breaking the container. We had to thin the bananas and rebuild the brickwork.

Over time, we gradually replanted vegetation in the front yard and courtyard, including several fruit and nut trees. There was a large pine tree in a very tight spot by the corner of the house. Severely malformed, it grew at a 30-degree angle into the neighbor's yard. We don't ordinarily approve of cutting down trees, but we made an exception in this case because we saw no way to right the tree. So we felled, split, and dried it, and eventually it became firewood.

As we cleaned up around the back of the house, we discovered that the porch area had a patio of bricks that had been completely covered with dead grass and soil.

The back yard was mostly bare except for some grasses in the lower area. There was only one tree, a grapefruit, growing in the back yard, and we often wished aloud that we'd inherited *any* other kind of fruit tree. After all, grapefruits were way down on our list of desirable fruits. And yet a couple of years after moving here, we finally juiced the tree's fruit and found ourselves enjoying one of the best juices we've ever had.

In the more than fifteen years since we have moved here, we've planted many more trees in the back yard.

One of our first building projects out back was a compost pit and worm farm, because we don't believe in tossing out recyclable garbage for the city workers to take to the dump. We also made a vegetable garden, and a coop and yard for our chickens and potbellied pig. We placed our rabbit hutch directly over the compost pit so that urine and droppings continued to feed the earthworms, and we rarely had to clean out the hutch.

We regard our small urban homestead as a research station. Here we are able to try out many gardening, recycling, and building ideas to see if they will really work or if they need refining. We have endeavored to let our living home laboratory be truly an extension of our values and our thinking.

We also realize that one's home is much more than the physical structure and that one can improve upon and elevate mundane physicality by thinking more broadly and more clearly. This we strive to do in little ways and big ways, as we'll relate in the chapters that follow. We are aware that *how* we do things is as important (if not more important) than *what* we do.

When we have the income to do so, we sometimes get outside assistance to do various projects, as when Mike Butler installed our solar water heater. But we are also very interested in doing whatever we can ourselves, using recycled materials, for as little money as possible, in as natural a way as possible.

This is an ongoing adventure and experiment in living lightly on the earth—in the city. We'll tell you how we proceeded, step by step. Please join us.

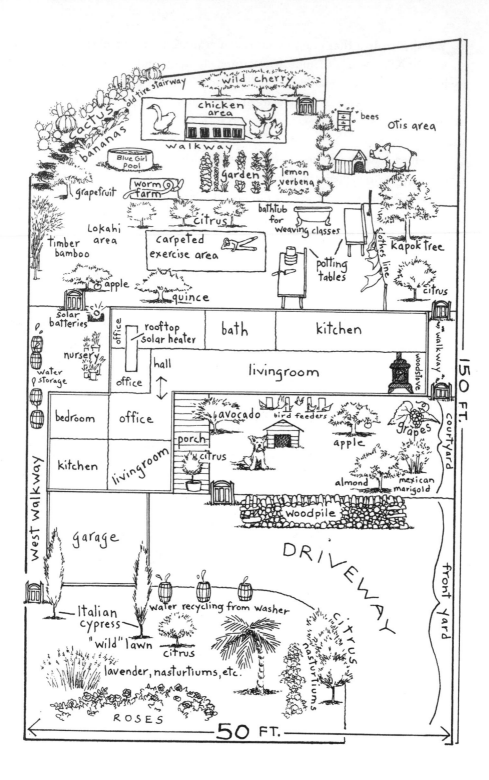

cactus
bananas
old tire stairway
wild cherry
chicken area
bees
Otis area
walkway
Blue Girl Pool
garden
lemon verbena
grapefruit
worm farm
citrus
bathtub for weaving classes
Lokahi area
timber bamboo
carpeted exercise area
potting tables
clothes line
kapok tree
citrus
apple
quince
solar batteries
rooftop solar heater
bath
kitchen
nursery
office
office
hall
livingroom
woodstove
walkway
water storage
office
bedroom
office
avocado
bird feeders
grapes
courtyard
kitchen
livingroom
porch
citrus
apple
almond
mexican marigold
Woodpile
West walkway
garage
DRIVEWAY
front yard
Italian cypress
water recycling from washer
citrus
nasturtiums
"wild" lawn
citrus
lavender, nasturtiums, etc.
ROSES
50 FT.
150 FT.

6

# 2  The Yard

*A nation that destroys its soil stuff destroys its soul stuff.*

—VERNON

When we first moved into our home, the front yard was ugly—barren and oily.

Out front, the previous residents had used the yard to park their cars. It is about 35 feet by 15 feet. Just a bit of crabgrass grew around the edges. The inner front yard, which we now call the courtyard, was almost as barren, though there were a few trees there.

One of our first improvements, once we had removed bits of old metal, wood scraps, logs, and an old shack, was to very heavily mulch the barren yard and the neglected courtyard areas. Mulch consists of natural materials such as wood chips, leaves, grass clippings—organic matter that can be spread on the ground to hold in moisture. As the mulch decomposes, it helps to increase the soil's fertility.

While driving home one day, we saw a yard that was covered with fall leaves. We had our rakes and bags with us, so we pulled over and knocked on the door.

"May we rake up your front yard and take the leaves with us?" we asked the elderly man who came to the door.

He was silent for a moment, uncertain what we had said, or perhaps suspicious of our intentions. We repeated our request.

"We'd like to rake up your yard. We don't want to charge you. We just want the leaves to use for mulch."

By now his wife had come to the door and we had to repeat the request again. They seemed to realize that we were sincere, and agreed.

As we raked, they began to laugh at their good fortune with sheepish smiles—someone had actually knocked on their door requesting to do something for free that they usually had to pay for.

"Take all you want!" the man told us, cheerfully and loudly.

We busied ourselves filling up about four large trash bags of the yellow leaves, and they watched us from their window with large grins. We laughed to ourselves, too, and wondered if they would be telling and retelling this curious story to their friends and grandchildren.

When we got home, we scattered all those leaves around the needy front and courtyard areas. We knew that we'd have to add more and more organic matter before the soil would be fertile enough to grow plants, so we collected leaves from other sources as well and spread them in our yard.

Neighbors watched our leaf mulch project curiously. One weekend we had a strong windstorm and all the added leaves blew into our neighbor's yard. Not wanting to confirm any suspicions that our project was going to be a problem, we quickly grabbed our leaf blower and blew all the leaves back home.

We contacted an acquaintance who runs a tree-pruning service. This man and his crew prune trees and then chip up the prunings, and when their dump truck is full of chips, they take it to the local landfill and pay to unload the chips. In response to our invitation, they were happy to bring one load to our place instead and dump it in a huge pile onto our front yard.

The huge pile covered most of the front yard, and the central peak was nearly five feet tall. We knew the pile would get smaller over time as the chips decomposed. In fact, the pile had sunk down about a foot after the first week, and we spread the chips out on each side so we'd have a mulch that uniformly covered the entire area.

If you've ever been around a big compost pile, you know how it generates lots of heat as the contents decompose. We noticed our pile steaming in about two weeks, and we also watered it a few times to help the decomposition process.

One morning, a neighbor from next door yelled, "Your front yard's on fire!"

We ran out expecting to see flames somewhere but saw only the steaming chip pile. We reassured our neighbor that everything was fine.

In two years, after two big truckloads of wood chips, we were able to sink our hand down into the soil in the front yard, and wild plants had begun to thrive.

We added four citrus trees in front, a line of rosebushes bordering the sidewalk, and some herbs including lavender, epazote, and white sage. In time, other volunteer plants found this a likable environment, and soon the front yard was carpeted with a miscellany of nasturtium, tradescantia, and Peruvian mint.

The nasturtium actually washed down the hill from our northern neighbor's property, and the plants have thrived in our front yard, re-seeding themselves year after year. This was a very welcome "volunteer," because all tender parts of nasturtiums, including the flowers, are delicious in soups and salads. The tradescantia (also called "wandering Jew") seemed to appear on its own and thrived, as it tends to do, wherever it takes root. The Peruvian mint came from some of the cuttings we were then cultivating and selling at local farmers' markets. Though not a true mint, it is a succulent that can be used as you'd use regular mint, plus it gives off a pleasant fragrance if you gently press a leaf.

The two of us have talked about the problem of "lawns" many times. In some neighborhoods, a perfect green lawn seems to be a requirement, and the neighborhood association will jump all over you if you allow your lawn to fall into a more natural state of "disrepair," claiming that your sloppy grass will lower the property values or attract pests.

We've heard it said that the idea of a lawn harks back to our ancestral homelands in Africa, Asia, and the Great Plains, where fires and grazing animals created great grasslands with random clusters of trees. Is that explanation supposed to mean that urban folks need a tiny, postage-stamp-size savanna beside their homes, as if to symbolize a carefree, primeval existence? It's an interesting theory, but think of all the labor, tools, gasoline, water, and fertilizers that urban dwellers use to maintain that little patch of green. And for what? Unless we actually play lawn games on the grass with our families, we do all that work, mow the lawn, then typically throw away the "crop." Yet another of the many urban counterproductive activities that millions of us consider normal.

This is unfortunate, especially in the drier areas of the United States.

We have seen vast fields that once were desert—with all the cacti, snakes and reptiles, and wildflowers characteristic of unique desert ecosystems—leveled by developers. The land is bulldozed, hundreds of clone houses are installed, and every one of these has that patch of suburban green that is so totally out of place in a desert environment. Water, and more water, is required to maintain that pointless patch of green. As Pogo would say, "We have met the enemy and he is us."

We cringe every time we see the television commercials for lawn herbicides that are guaranteed to kill just about anything that lives. These ads show a gardener using poisons to kill dandelions in lawns—but dandelions are one of nature's most nutritious plants! Whatever you do, don't kill your dandelions. Leave them alone, or learn how to use them.

And did you know that the annual amount of energy, water, and fertilizer required for the average acre of lawn is more than double what is required for an acre of corn? In addition, there is the pollution caused by runoff containing lawn fertilizers going into the water table. Some of the chemical treatments designed to "green up" your lawn contain chemicals such as 2-4-D, which has been proved to cause illness in people. And some people even plant "organic" vegetable gardens that are surrounded by toxic lawns!

Far better for all involved, and for the planet, is to discover the principles that govern all facets of the natural world and then strive to live our lives in harmony with those principles. Considering the high cost of food and fuel, as well as the problems of overflowing landfills, water shortages, and the harmful side effects of herbicides and lawn mower exhaust, why don't we do away with superfluous lawns and convert these areas to productive uses, like the Victory Gardens of World War II? That is the decision we made, and we have never regretted it.

We're also glad that we don't live in the sort of neighborhood where tidy grass lawns are expected. To some eyes our yard would probably appear disheveled, though we see it as filled with a diversity of active living things.

During our early years here, we planted vegetables on the "parkway," that narrow strip of ground between the sidewalk and the street. One year, we had corn, tomatoes, New Zealand spinach, squash, and herbs growing out there. Several neighbors came to help in the preparation of the soil and the planting, and that project had the great additional benefit of bringing neighbors together.

We didn't try to maintain an edible garden out there for long, though we do still grow some herbs in that section of the yard, because many dogs walked by on our street and often found our planted areas to be suitable toilets. Great for the dogs, not so great for us. Also, the plants out on the parkway are right next to the street and therefore exposed to pollution from cars.

Our front yard now looks like a jungle of sorts, since a palm tree has grown up and we allow some castor plants to grow. Castors are poisonous, so they are never eaten; the poison is the most concentrated in the seeds but is found in all parts of the plant. Yet in spite of that, the plants do have some uncommonly good features. They grow quickly, producing very large palmate leaves. This means that they will

CHRISTOPHER NYERGES

*Dolores with produce from our experimental neighborhood parkway garden.*

quickly provide a privacy screen and shade. And where they are growing thickly, it feels like a humid jungle when you are underneath these tropical plants. A large, thick patch of them would actually trap some methane (from decomposing plants and from mulch in the soil) and would therefore create conditions that are good for other plants.

This yard also has countless earthworms, rich soil, and supports a host of birds—especially sparrows and jays—who feed on the various seeds and insects.

We don't want to imply that such a yard is care-free, because we do need to water occasionally, and we need to thin and trim certain plants. But we are not fighting nature. We are allowing a miniature wild environment to exist right around our house. No poisons are ever needed, and—because of the thick mulch—only minimal watering is necessary. Once an area has been heavily mulched, we do not generally need to add more mulch for a year or more. (If, however, you decide to maintain a lawn, remember the advantages of leaving grass clippings and other trimmings on the lawn after mowing instead of sweeping them up and hauling them away, which in many cities involves additional expenses and fees. Grass thrives when it's allowed to grow a bit longer and is then replenished by the organic matter in clippings.)

Though the extent of wildness that we prefer might not be for everyone, it seems indisputable that the surroundings of most urban and suburban homes could be put to better use. Even without letting your yard go completely wild, it is relatively easy to plant several fruit trees (dwarfs for limited areas) in your front yard and create a small orchard. We recommend that you find out what grows best in your climate, in your part of the country, and grow those plants.

If you insist on the more wide-open, expansive lawn effect, consider some appealing alternatives to conventional grass. For example, owners of English estates who had large expanses of lawn to maintain found that they could plant low-growing perennials that required very little upkeep. The possibilities include plants such as red clover, creeping thyme, mint, and dwarf chamomile, all of which can be planted just about anywhere, and which are fragrant and tasty. Other plants that do well as a lawn substitute—and which also provide food—include New Zealand spinach, wood sorrel, Madeira vine, purslane, cresses, and many others.

It is hard to believe that our front yard area was once hard-packed adobe partly covered with scraggly crabgrass. Today it is not only wildly beautiful but provides a wonderful home for neighborhood birds and much-needed oxygen for all of us.

The sheltered part of our front yard—the area within the L of the house, which we call the courtyard—was not quite as barren as the front but had long been untended and ignored. We mulched the courtyard soil with bags and bags of grass clippings, sawdust, wood chips, and straw, all acquired for free or almost free from others who didn't see their value and were glad to get rid of these organically rich materials. We removed the banana trees that were poorly located there, because the area was really too small to accommodate the ever spreading roots of bananas, and planted a wonderful avocado in their place. The avocado turned out to be a perfect choice. It eventually provided shade and privacy for our court-yard area and gives us an annual, top-quality crop of oil-rich avocados.

This courtyard was open to the driveway and therefore the street, and we immediately built a fence around it to provide privacy and secu-rity. We planted grapes on the fence and stacked our firewood along the outside.

Also in the courtyard we planted a nonpareil almond tree, a tanger-ine, an apple, an Asian pear, a plum, and a grapevine. The pear and plum eventually died, because of either gophers or our dogs, but the other trees have done well.

Before we had three dogs, the courtyard produced an abundant crop of lamb's-quarters. This annual didn't survive in that area with the dogs' constant running, digging, and rolling, but we still have lamb's-quarters in most of the other areas of our yard.

Lamb's-quarters is a relative of spinach, but even more nutritious. It is a common wild plant worldwide. Though we never planted it, we en-couraged it. We used the greens in soups and salads, and the seeds we've added to both soups and bread batter. We harvested food for many meals from that patch. One year there was a very healthy lamb's-quarters plant that grew up just outside our entry and eventually reached a height of 12 feet. We are certain that that is either a world record or very close to it. The plant amazed everyone who came to our place. We let it go to seed and would watch a dozen or so sparrows feed off it in the mornings.

At one time, we hung a beautiful old lamp from a tall tree in the courtyard and set up a rustic outdoor table with tree-stump chairs. We had many gatherings and meetings out there, in the day and at night. It was a very pleasant place to sit and talk.

But we now keep two dogs out there, and that yard has become their place (we have dismantled the outside table and have our meetings else-where).

## THE BACK YARD

When we moved in, our back yard was as dry, unkempt, and neglected as any yard could be.

This area has two levels: the entire back yard is 50 feet wide by about 40 feet deep. The northern half is higher by four or five feet, with a retaining wall holding the embankment in place. The lower section (that is, the southern section closest to the house) is at house floor level. There were some half-dead, overgrown pittosporum bushes off along the east wall and one tall grapefruit tree along the northwest border. Otherwise, the entire back area appeared almost hopeless—infertile soil and brittle grasses. It was obvious that no one had ever made any attempt to clean, improve, maintain, or irrigate the yard in the previous ten years.

One of our first tasks was to designate an area just east of the grapefruit tree as our compost pit and worm farm. We used old railroad ties and created an area about five feet square. This afforded us enough space to add new yard trimmings and kitchen compost, yet have it large enough so that the redworms (a fast-breeding type of earthworm) could survive.

We also needed to quickly construct a chicken coop (which we enlarged a few years later), because we'd brought a flock of forty-eight chickens with us, leghorns and Rhode Islands reds. Our first chicken coop was a large, old wooden box we found on the premises—that was where they roosted—and their "run" was a simple arrangement of stakes and chicken wire, about eight feet square.

Our back yard had been overgrown. It was so overgrown with dead, untended grasses that it took us half a workday to realize that the "floor" of the back patio area had been laid with red bricks! What a surprise.

We had our work cut out for us, considering how our new surroundings had been neglected by the previous residents. We sat down and drew maps for the yard areas and made plans for all the trees we wanted to plant, and the garden, and the rabbit coop, and the nursery, and the exercise area. And then we resolved that we'd always try to make a notable improvement each time we worked on the yard, house, or garage. Our overall concern was that we create a home where we could put into practice and demonstrate the principles of ecological living. We didn't want to create extra, unnecessary work for ourselves, but we were determined to save rainwater, recycle all our own garbage, live energy-efficiently, produce at least some of our own food, have a healthful environment to do exercises, and enjoy a peaceful place to work and think—not too much to ask, surely.

We also feel that it's important to look for the sacred in everything and to build memories into projects. For example, our front gate and parts of our chicken coop came from wood recycled from the martial arts studio of Barton Boehm, a friend of ours and a master of the Sei-Ken ("kind hand") system.

And each time a pet dies, we bury him or her under a bush or tree, and thereafter that growing plant is a living memorial to our beloved animal companion.

## TREES: THE REAL SIGN OF WEALTH

One of the easiest and best long-term investments you can make toward improving the earth and the ecology of your own yard is to plant trees. The late Egyptian president Anwar Sadat knew the great value of trees. He once mandated that all adult Egyptian males plant one fruit tree per year.

Trees are nature's air purifiers. On the underside of every leaf are hundreds of thousands of cilia (hairs) and stomata (pores) that readily capture free-floating pollutants and gases and pass them to other parts of the tree to be metabolized. Trees with their dense, overlapping foliage also provide a measure of soundproofing around your home, which is especially critical in the city. Trees hold the earth together with root systems that are usually at least as large as the aboveground part of the tree. They provide shade, homes for birds, and loads of oxygen, that sweet, invisible stuff that sustains human life.

Trees also lessen the need for mechanically cooling a house. The tree's shade and transpired moisture cool the air that circulates around and through a home's open windows.

Although any tree is beneficial, on our little urban homestead we always select a tree or shrub that will offer a diversity of benefits: nutrition, medicine, utility, and fragrance. Some trees—the lemon, for instance—provide all of these. Because we live in a region where water shortages are chronic, we also select trees on the basis of drought tolerance. And we appreciate the ability of certain species of tree or bush to create a quick fence.

Based on research and conversations with others, we keep a list of trees with the qualities we have determined to be most important, then plant those trees that rank highest on our list. One of our best decisions was to plant an avocado tree more than ten years ago, when we moved here. Today that avocado provides us with foliage for privacy, abundant high-quality fruit, and top-quality leaf mulch.

Another good choice is citrus trees. When we moved in, we already had that huge grapefruit tree out back, which produces excellent juice. In general citrus trees are drought-tolerant, and they yield healthful fruit and juice as well as a rich fragrance in the spring. They can also be used as the basis for unique products, such as the lemon "soap" we make from peels. Since arriving here we've planted oranges, lemons, kumquats, tangerines, and even more grapefruit. These trees are planted all over, some in the front, one in the courtyard, in a line out back, and so on—not all in one place.

If you wish to turn your yard into an urban orchard, begin by making a list of those fruits and nuts that you most enjoy. Then consider the soil and light requirements of each; consult reference books and pose questions to local botanists, gardeners, and nursery people, who tend to enjoy sharing their knowledge of plants. Compare your "wish list" with those varieties that are known to do well in your part of the country. For us, the plants that thrive in our coastal desert plain include mission figs, loquats, sapote, natal plum, Gordon apple, Santa Rosa plum, most apricots, some bananas, most avocados, mission olives, and many others.

Oaks are especially worth planting at home. In terms of a food crop, the acorns can be readily leached of their bitterness and used in bread recipes. But the most remarkable characteristic of oak trees is their high transpiration rate: an oak tree will transpire up to 300 gallons of pure water per day, through its leaves, into the environment where it is planted. A big redwood is believed to transpire up to 500 gallons of water a day into the air through its leaves! That's the magic that you participate in when you plant and nurture trees.

Remember that trees create their own microclimates; they reduce the speed of the wind across the land; their roots actually raise the level of the local water table; and their presence increases the population of worms, which increase the fertility of the soil. In addition, shade from trees near your home can reduce your summer utility bills.

Research by the National Wildlife Federation has shown that forested land returns ten times the moisture to the atmosphere as deforested land. Clear-cutting forests is a form of national suicide. According to Dr. Richard St. Barbe Baker, also known as "The Man of the Trees": "You can gauge a country's wealth—it's real wealth and health—by its tree cover. . . . We forget that we owe our existence to the presence of trees."

A solution to ecological devastation is to plant trees, starting at home. You'll increase the fragrance and beauty of your yard and get delicious, fresh, natural fruits and nuts in the process. Wherever you live, when you plant and care for trees, you become part of the solution!

CHRISTOPHER NYERGES

Left: Freshly picked prickly pear cactus fruit.
Below right: Eucalyptus leaves and pods. Tea from the leaves relieves congestion; powder from the pods can help heal minor cuts.
Below left: Lamb's-quarter. We use the leaves in salads and soups and bake the seeds into bread.

CHRISTOPHER NYERGES

CHRISTOPHER NYERGES

An autumn scene from our back yard wild food cooking class—acorns in molcajete and a pumpkin.

CHRISTOPHER NYERGES

# 3 Homegrown Foods

*Once we stop fighting and start thinking, we discover that the enemy may be a friend.*

—MARGARET PACSU, on David Suzuki's
"Nature of Things," regarding water hyacinth

We are convinced that a big source of problems for urban dwellers is that they don't think like farmers; they don't even believe that they *can* or *should* think like farmers. Sadly, most contemporary farm families don't think like traditional farmers, either.

In our view, "thinking like a farmer" means looking for every opportunity to use your own land (even if it's a small urban lot) to produce as much of the food, medicine, energy resources, building supplies, and other useful materials that it possibly can. We ourselves see the land as a primary source of sustenance, whether that land is rural acreage or an urban or suburban yard. There is no good reason why urban dwellers cannot produce at least some of their food.

As noted in the previous chapter, we undertook research and planning before we began to purchase or otherwise obtain the plants for our yard.

First, we compiled a list of fruit trees and other useful plants that are most successfully grown in our geographical area and climate. Our list included vegetables, fruit trees, bamboo, bananas, wild plants, trees that yield some useful materials (such as kapok), and so forth. Then we made a list of the fruits and vegetables that we enjoy most and a list of the raw commodities that we've determined we need or would like to have. We began to see where these two lists overlapped, and that overlapping became our list of top priorities. In time, we learned that many of our top choices were not easy to come by; some we have never obtained to this day. For instance, we've heard of a wonderful, banana-like plant called ensete—alas, you can get it only in Ethiopia.

As an example of one of our priority plantings, we have determined that citrus trees are almost uniquely suitable for our region. They are drought-tolerant, and they combine food, beauty, and medicine all in one. Their fragrance is wonderful in the spring when they flower, producing a heavy sweetness that attracts bees from all over. Thus we have planted many citrus trees, including kumquats, tangerines, various lemons, various oranges, and grapefruits.

Citrus do so well on their own in this climate that we have found surviving citrus trees that were planted more than seventy years ago in the Angeles National Forest, surviving on sites where there once were cabins that were later burned out or flooded away. Anytime we encounter plants like that, we know we are dealing with strong stock: survivors.

Citrus fruits make good food and juices, and also great medicines. They contain vitamin C, bioflavinoids, and other health-promoting nutrients and minerals.

Today's typical "lemonade" is an artificially flavored beverage loaded with white sugar, just another junk food; but the original lemonade, sweetened with honey, was just the thing for enabling your body to deal with the stresses of hot temperatures and heavy labor. If you add just a little molasses to this drink, you are truly drinking your needed vitamins and minerals.

When the peels of lemons are soaked in water, they begin to soften into a gel-like substance that has many uses. We have rubbed it into wood as a finish, rubbed it onto our skin over sore muscles, and even mixed it in an electric blender to make a fertilizer for citrus trees.

You may recall from the previous chapter that the only tree in our back yard when we moved in was a large, old grapefruit. We had tried eating the fruit for breakfast and found it too tart and bitter. For the first year or so, we grumbled to ourselves that it was too bad there was such a "worthless" tree here.

But one day we decided to try juicing the fruit, and we were dumb-founded to discover how remarkably smooth and delicious the juice was; we also found that drinking it when sick helped us quickly recover. We called this juice our liquid gold, and we came to call the tree Great Grapefruit. That just goes to show that the thing you need is often right before you, though perhaps "disguised."

Along the very front of the property, we also planted seventeen different roses, all selected because of their fragrance, a quality that is lacking in far too many modern roses. A few other roses were planted in the courtyard and back yard. In addition to enjoying their scents, we eat the petals in our salads, and the hips (fruits) can be collected and eaten, as they're one of the richest natural sources of vitamin C. Rose hips are commonly used in teas but can also be made into soups, jams, and jellies.

Apples were high on our wish list, too, but unfortunately there are few apples that will consistently do well here in this Mediterranean-type climate. We planted an apple variety called Gordon. The Gordon is a wonderfully large tree, and though we eat the fruit, these apples seem best for applesauce and juice—a rich, sweet, tangy juice that always seem to remind us of some distant childhood memory. Perhaps the taste evokes a memory of foods that actually had flavor, unlike the majority of produce in a typical American market today.

We also planted an Anna, perhaps our all-time favorite apple. The fruit is tart, crisp, and delicately flavored. The tree is a dwarf, and we like it so much we are trying to root cuttings from it so we can have a few more. We also have a pear—related to the apple, and also on our priority list—but the variety we chose has never grown very well here. We had an Asian pear in the back yard, but the dogs killed it by chewing on the trunk. We planted what we thought was another Asian pear, but it had been mislabeled by the nursery and this tree turned out to be a quince. According to *The Whole Foods Companion,* quinces are actually in the rose family, and although they are often used as rootstock for pears, "the two fruits simply cannot be hybridized." We have not yet figured out what to do with quinces, but we are considering grafting apple branches onto the tree.

In our courtyard, we planted a nonpareil almond tree, because almonds are not only good food but also good medicine, as almond oil is a demulcent and emollient. After ten years, it's a big tree, fragrant in the spring with its white flowers. Almonds you pick yourself don't look like those you buy in the store, the rinds of which have been removed. With their rinds still on, our homegrown almonds somewhat resemble dried prunes, and they have a better flavor than store-bought types. We do have to pick them when they ripen or the neighborhood squirrels and birds will eat them. One year, the squirrels ate nearly every single one!

From the standpoint of readers living in many parts of the country, avocados seem very exotic, but they grow easily in California, without the need for spraying or high-maintenance care. One of our first plantings when we moved here was the Whitsell variety of avocado, which we planted in the courtyard. It has grown up tall, providing the shade and privacy we needed, and it produces a good annual crop of meaty, oily avocados. We have heard of many recipes that can be made with avocados, though we generally just eat them in salad or as a side dish.

We have all heard that avocados can be "fattening," but they're also very nutritious. According to the USDA, avocados are rich in fat and calories (up to 589 calories and 58 grams of fat, per pound). A pound of avocados also contains up to 2,082 milligrams of potassium, up to 1,000 milligrams of vitamin A, up to 145 milligrams of phosphorus, 34 milligrams of calcium, 14 milligrams of iron, and lesser amounts of ascorbic acid, niacin, riboflavin, and thiamin.

We know of at least two individuals who claim to have survived for months primarily on avocados (in one case, the man had citrus fruit as well as avocados). While we do not plan to live on avocados exclusively, we remember those stories just in case we *might* someday have little other food but the avocados.


## THE BACK YARD

Back yards seem to be a disappearing aspect of Americana. Since the 1950s, homeowners increasingly viewed back yards as "unused capital" and paved them over, subdivided them, and built on them. What a shame, for once gone, yards are nearly always "gone for good."

We can walk out our back door to greet Otis (our potbellied pig). We can look north to a virtual forest of green as we sit out back doing our morning meditations and planning the day. Almost every morning we witness hummingbirds feeding, squirrels scampering in the treetops, and we inhale the sweet fragrance of oxygen and flowers. Such a back yard is like a miniature Garden of Eden. Wherever possible, we should resist the serpent's temptation to destroy these mini–life-enhancing pockets of urban wildness.

## Apricots

Over the years we have planted many apricots, collecting the large seeds from the trees at Dolores's mother's house and sprouting them. Apricots require no grafting in order to get fruit. They are relatively easy to sprout and grow, and they get quite large as they mature. We have only one apricot tree planted in the ground; the others are in large pots. We have successfully replanted the seeds of our neighbor's drought-tolerant apricot, a variety that he calls his Merced apricot because he buried a dog by that name under the tree.

## Bananas

Bananas are a readily digestible, flavorful food that provides many necessary vitamins and minerals. Granted, it isn't possible to grow bananas in all parts of the United States, but they certainly do grow here. There are eight varieties that are known to do well in our region. We have two small patches of good bananas for our area.

Bananas available at the supermarket and bananas you grow yourself are wholly separate entities. We've had some especially tasty bananas grown by our friend Richard Buckland, who is a longtime member of the Rare Fruit Society here in Southern California. Once while visiting him, he served us some small, fresh bananas that he had picked a few days earlier. They were meaty (not mushy), richly sweet, and flavorful, not bland. They tasted like a substantial food, not just a snack. By contrast, the bananas we buy in the store are picked and shipped green, then gassed with ethylene to forcibly ripen them, and they never reach their peak of flavor, nutrition, or sweetness.

There are actually hundreds of varieties of bananas, the majority found only in the tropics. Tall plants, dwarf plants, with big fruits or little fruits, and a great variety of fruit color, shape, quantity, and quality. Fruits can be round or short, long or skinny, and red, yellow, or green (and all shades in between), as well as hard, soft, sweet, or starchy. Some are crisp, and some very mushy. The so-called common banana is Gros Michel, often referred to as Bluefield (*Musa paradisiaca-seminifera*). The banana has this Latin name because the "fruit" was believed to be the fruit of paradise (the one with which Eve tempted Adam). This belief began during the Crusades.

How do you cook a banana? Drop it in boiling water, unpeeled. Simmer twenty-five minutes, and season as you wish. Or bake for forty-five minutes in a 250-degree oven, unpeeled.

Bananas can also be added to vegetable dishes or dessert items, sautéed with honey, or dried for a nutritious, high-energy food. It is far more economical to dry your own in a food dehydrator than to buy dried bananas. They can also be dried by stringing cut slices on a line that is then suspended in the sun or in a car parked in the sun.

Keep in mind that "tests" of the nutritional qualities of bananas consider only the fruit, not the skin. There is a simple way to get a more complete food from each banana. After eating the ripe fruit, take your spoon and scrape the inner peel and eat that, too. It is very "meaty" and a good complement to the fruit.

Few edible bananas are truly wild. As a rule they need the nurturing of humans to produce reliably good fruit. As for how to grow them, it's important to understand that bananas are not *trees*. They are actually perennial herbs—nonwoody monocots, actually—not at all like fruit trees.

Each stalk fruits once and then dies, so it's best to cut down the fruiting stalk after the fruit ripens so that new shoots surrounding the main stalk will grow up and produce. Bananas produce best if well-watered and fertilized (we've found that pig manure is good for bananas). A frost will cause a severe die-back but generally does not kill the plant.

There are at least fifty medicinal uses for every part of the banana fruit and plant, according to *Medicinal Plants of India and Pakistan*, by J. F. Dastur, F.N.I. For example, from the pounded peelings of ripe bananas you can make a poultice, which can be applied to wounds to inhibit infection and promote healing.

Banana skin has been praised for being a "bacteria-proof wrapper." The skins of ripe bananas contain powerful antibiotic substances that are effective against disease-causing fungi and pathogenic bacteria. We found details on this in an article called "Antibiotics That Come from Plants" in *Yearbook of Agriculture*, 1950–51.

And when you are done eating this wonderful fruit, take what's left of the skin and polish your leather boots and shoes. It really works! Rub the inside part of the peel energetically onto the leather and massage it in. Whatever skins we have left over, we feed to Otis, our potbellied porcine-person (a.k.a. pig).

## Bamboo

When we moved here, we were given some roots of timber bamboo, the giant bamboo widely used in Asia to make scaffolding for high-rise construction as well as in hundreds of other applications, including temple and ramada frames, decks, baskets, fishing rigs, cups and bowls, and much more.

*A section of timber*

We planted two roots, hoping to cultivate a simple privacy screen. Today we have a beautiful and very tall "wall" of bamboo along the west side of our back yard. We have cut stalks from time to time to make things such as chicken perches, fish spears, "fire pistons" for making fires, containers, garden trellises, gates, sprinkler supports, digging implements, sculptures, musical instruments, and more.

Bamboo is truly a multipurpose plant. We figure that in a few years we'll have enough tall stalks to build a three-story ramada, as the mature stalks are quite tall, about eight inches thick, and exceptionally strong under vertical compression.

The atmosphere beneath our bamboo wall is wonderful—cool in the summer and shady, with many birds making their nests in the upper reaches of the stalks.

At night when the sky is clear, we stand under the towering bamboo and hear drops of water fall from the leaves. The sound is like a gentle rain, an effect known as "tree dew" or "tree rain." We particularly enjoy this late at night when the surrounding neighborhood is quiet. We take our dog Lulu out to use her toilet area, and we stand there listening to the bamboo dewdrops. This is always an uplifting and wonderfully subtle experience. Often Lulu will lift up her nose and sniff the air, listening to the drops gently falling on us. And often, way in the background, we'll hear Blue Girl . . . snoring. We really don't know for certain if geese "snore," but this is a muffled sound, very similar to a human snoring, and we try to be very quiet so we don't disturb her.

You have also no doubt heard of eating bamboo sprouts, but edible bamboo sprouts come from a different variety than the bamboo in our yard. We sampled one of our "sprouts" (raw and cooked) and it was very tough, very bitter, and unappealing as food.

### Nursery

We have an area on the side of our house where we maintain a small nursery. We occasionally sell potted plants, wholesale or retail; nearly all of these are unique plants that we have propagated at home.

Our nursery includes such plants as aloe vera (good for skin conditions and general skin care), Peruvian mint (a succulent and aromatic ornamental), and various cacti. We especially enjoy growing edible prickly pears, Chinese ginger, carob trees, citrus trees, various onions, fig trees, and fragrant geraniums.

When we sell plants from our nursery, we hope to achieve more than merely making money. For instance, we want the selection of plants we are growing to reflect our overall way of thinking: every plant must be useful, edible, medicinal, or fragrant, ideally with as many of these positive attributes as possible. The plants we grow for market should also be appropriate to our environment and be drought-tolerant.

Thus, many of the plants we propagate for sale are the plants that we grow around our yard for our own use. Often we collect seeds or trimmings from our own plants to create the plants we sell.

For instance, our prickly pear plants are abundant, and when we have to thin a section of one of these that is getting large, we cut off pads and root them. Soon we'll be selling the best of our varieties to others. We also collect carob pods every year and toss the seeds into a large pot. Most of those seeds sprout, which we then repot and sell as trees. We do the same with citrus seeds from our grapefruit tree and at least one of the other citrus trees.

Some of the plants we sell are grown from cuttings from trees or vines in our yard or from plants that we've found in the local mountains, including figs, apples, and grapes. The cuttings of each of these plants are readily rooted and grown into vegetative clones.

We have no interest in selling plants that are fashionable only during one year or that are sold merely for looks, like so many of the ornamentals you see at commercial nurseries. We feel that these living floral beings should be carefully selected for as many beneficial traits as possible.

## Seeds

When we've decided that we want to plant a particular variety of tree, bush, or vegetable, we will often research the available varieties until we find a source of seeds or living plants that are particularly well suited to our location. At other times we propagate new plants from those we already have, saving seeds from favorite plants in order to cultivate a next generation.

We have often been surprised to read in garden magazines that if you plant citrus seeds, you will not get the same tree as the parent, but that has not been our experience. We have saved our citrus seeds for years, planted them carefully in hospitable spots, and grown numerous trees. Our neighbor has an old-variety lemon that we have always liked—he calls it "Luke"—and so we planted the seeds from his lemons and have grown several more trees. This is the second year our seedling trees have produced lemons, and they are just like the fruit from our neighbor's tree—we call them "Luke, Jrs."—tangy lemons unlike any we have ever found in grocery stores.

According to Dr. James Bauml, botanist at the Los Angeles County Arboretum in Arcadia, California, citrus seeds are polyembryonic, meaning that there can be more than one seedling per seed. Of all the seedlings that sprout from each seed, one will be a sexual (fertilized) embryo, which will not produce the same fruit as the parent. However, the fruits of all the other resulting seedlings will be identical to the parent. So if two seedlings arise from a seed, one will be the same as the parent and one won't. If five seedlings arise from a seed, four of the seedlings will be identical to the parent. The percentage of seedlings that will be identical to the parent is a function of how many seedlings eventually sprout from each seed.

Today, we sprout mostly lemons and grapefruits, though using the same methods we have sprouted other citrus as well, including kumquats, oranges, tangerines, and limes.

Dr. Bauml states that the best time to sprout citrus seeds is around the time when the fruit from that tree is naturally ripening. That makes sense to us, and it's convenient, because this is the time when you are picking the fruit and need to deal with the seeds. Unlike other seeds, which might not be planted until the following year, citrus seeds shouldn't be dried and kept for later use. Dr. Bauml explains that citrus seeds are ephemeral and should be planted as quickly as possible.

Because we have a large grapefruit tree whose fruit produces delicious juice, we save and plant the seeds from that one. We begin by planting up to twenty seeds in a 1-gallon pot, or approximately a hundred seeds in a large 5-gallon pot. We cover the seeds with fine compost and water them every few days. In less than a month, we observe that the majority of the seeds have sprouted. Within two months, we have a pot bursting with grapefruit sprouts about two inches tall.

Once the seedlings have grown to about three to four inches, we transplant them, putting one seedling into each 1-gallon pot. Our potting procedure involves first adding a few scraps of old clothes to the bottom of the pot, which helps to retain moisture and prevents soil from washing out of the bottom of the pot. Then we fill the pot with rich, wormy compost and place it on our potting bench where it can get mixed or indirect light each day. The young citrus trees will stay in this 1-gallon pot for another year or two, until they are big enough for a 5-gallon pot, at which point they are sold or planted directly into the ground.

We once made the mistake of transplanting all of our potted grapefruit sprouts under our "mama" grapefruit tree, thinking that the babies would like to be near the parent—hey! we're thinking like humans, aren't we?—and that they would like the shade. Though that reasoning may have been sound, our goose, Blue Girl, managed to get to the place under the big grapefruit tree, and she picked out and killed most of the sprouts. Now we keep the grapefruit sprouts in another area. Sprouts need to be protected from animals and birds who will find them tender and inviting.

Citrus trees are easy to care for here in Southern California, even in small city lots. They like water and rich soil, and they do well in full sun.

Once established, they are quite drought-tolerant. As we've already stated, we've seen citrus trees in the Angeles National Forest at locations where cabins were built long ago. Because these cabins were all destroyed no later than 1938, these citrus trees have had to survive at least sixty years on their own.

Citrus are perhaps the ideal tree for our area. In addition to the easy care and drought tolerance, they provide us with oxygen, fragrance, food, drink, and medicine. We plant citrus trees wherever there is space, and we never grow tired of them.

## Cuttings

In addition to seed saving and replanting, we have also produced new generations of plants by means of vegetative cuttings.

Many people do not realize that you don't need to go to a nursery and make a purchase every time you want another tree in your yard. There are several types of fruit trees whose cuttings readily root, and from rooted cuttings you can grow a tree that will produce fruit identical to that of the parent. This is a form a cloning—more precisely, a form of asexual reproduction.

Some of the plants whose cuttings readily root are figs, apples, grapes, blackberries, willows, roses, jujubes, and currants. Among those that do not root readily are avocados, peaches, nectarines, citrus fruits, and macadamias. We have had a high success rate with figs and a slightly lower success rate with apples, roses, and currants. You can learn what will and won't work through trial and error. You can save some time by consulting a garden manual that covers vegetative cuttings.

To prepare cuttings, first remove most of the leaves, keeping only a few on the very top so that photosynthesis can continue.

There are numerous ways to root the cuttings. For instance, you can place them in water, changing the water every week or so. Eventually the cuttings will develop many small, white rootlets, and you'll know that it's time to plant them. Willow cuttings root very well in water. Horticulturists have noted that willow adds some unidentified substance (perhaps salicin, though no one knows for sure) to the water, which actually helps other cuttings to root. You can try this and observe your results.

Though we've used the water method, we have found that we get the best results by putting our cuttings into either pure diatomaceous earth or vermiculite (both mixed with water), or into worm castings. We fill 1-gallon pots with whichever potting medium we are using, then insert up to six cuttings into each pot. The potted cuttings should remain in a shady area and be kept moist.

At the Los Angeles County Arboretum, the preferred medium for cuttings is a mixture of about 80 percent perlite and 20 percent peat moss. Sometimes sterile sand is used. Dr. James Bauml states that after all the leaves except a few at the top are removed, the cuttings are placed in 4-inch pots, often several to a pot. These potted cuttings are then placed on the arboretum's mist bench, where the cuttings are misted regularly until they begin to sprout. Sometimes cuttings will show obvious signs of growth within two to three weeks.

Timing is critical to success, because some cuttings will sprout only in certain seasons. In general, a good time to try to root cuttings is when the parent tree would normally be starting to produce new growth.

Dr. Bauml points out that an easy way to see if a cutting has rooted is to gently tug on it. If there is resistance, it is likely that it has developed many roots.

Once it is obvious that your cuttings have taken, you can repot individual cuttings in larger containers and let them continue to grow until they are large enough to be put into the ground. As for knowing when to transplant, this is entirely a question of how much space you have available; transplants can be anywhere from two to four feet in height.

Growing your own fruit trees and berry vines in this way is not difficult if you give the young plants the proper attention. Not only is this an economical way to increase your volume of plants and trees, but you can reproduce a favorite back yard fruit tree variety that cannot be purchased anywhere.

Here's an interesting historical note: Johnny Chapman, also known as Johnny Appleseed, the evangelist who wandered the Ohio River Valley, did not plant apple seeds. Chapman actually rooted apple cuttings, and—contrary to the myth—he did not wander aimlessly spreading his trees. In fact, he supported himself by selling the apple cuttings to homesteaders and farmers along the way. With so many schools strapped for cash these days, perhaps teachers could organize apple nurseries so that children can learn not only about plant propagation but also about conducting an honest and uplifting business.

Keep in mind that you must choose fruit tree varieties carefully, because some will not successfully set fruit in your climate. Merely because the tree is available at a local nursery does *not* mean that the tree is well suited to your particular environment. For example, apple varieties that do well here in Los Angeles are Gordon, Beverly Hills, and Anna. To find the specific trees and bushes that are best for you, check with an extension service agent or consult a garden book for your locale.

## WILD FOODS

How do you double the crops from your yard and gardens with no extra work? You learn to recognize the edible weeds and you eat them. Many people laugh when we offer this advice at our gardening classes, but once you begin to develop this skill it will alter the entire way you look at food production. In fact, some of our "gardens" are entirely wild, and most folks who glance at them would see only patches of weeds, not realizing that they were looking at a rich source of food.

For a long time we had heard about the "passive agriculture" practiced by many Native American peoples hundreds and even thousands of years ago. In general, this term refers to a way of working with the already present wild plants so that they become more fruitful. Areas would be "farmed," but there were no neat rows in square plots. Periodic burning was perhaps the most significant factor in passive agriculture. Burns removed dead grasses and dead wood, killed bugs, and laid down a layer of fertilizer in the form of ash.

While fire was sometimes used on a grand scale, small-scale actions included deliberate spreading of seeds, transplanting, scattering of manure, and careful pruning. Despite the intelligence and comprehension of natural processes reflected in these techniques, most people today would probably regard them as inferior to modern agriculture, pointing out that while passive methods definitely increase the size and volume of wild fruits and seeds, they require a much larger area of land to produce a given volume of food. It's true; maximum productivity per square foot of earth wasn't the measure of success for the native growers. They developed instead an agriculture that fostered the long-term, year-in and year-out fertility and fruitfulness of the land where they gathered foods and medicines.

We have gained many useful insights into what is meant by "passive agriculture" by carefully observing what happens when we collect and work with the wild plants from different locations in our yard.

For example, when we were actively selling wild greens at local farmers' markets, we intensively collected a variety of wild greens from the orchard and the extended back yard, an area that belongs to someone else who allowed us access. This provided us practical experience with the potential for actually "farming" (instead of merely "foraging") market greens on a very small piece of land.

One particular six-month period in 1995 was fairly representative of our yields. The annual California rainy season began in early January that year and we had record levels of rainfall, almost twice the normal amount. So we began intensive wild-food gathering for the various local farmers' markets we were attending.

Our label says "Wild Salad." In January, we were finding only chickweed, but this was okay, as chickweed alone makes a delicious salad when diced and dressed. As the season progressed, we were collecting the leaves and flowers of hedge mustard and common mustard, mallow, sour grass (also known as oxalis or wood sorrel), nasturtium, tradescantia, and the leaves of sow thistle and lamb's-quarters. We also collected the flowers of sweet alyssum. We were always careful to harvest by pinching or cutting, never uprooting, and we sold our greens the day they were collected or the next day. Whatever was left over we used in our own soups and salads.

At first, we went to three markets a week, and during the height of the season we were selling Wild Salad at five markets per week, none of them farther from home than a thirty-minute drive by car. We collected from three to five 5-gallon tubs of greens for each market. Our processing was minimal—no washing, just sorting to get the best greens. We packaged the wild greens in small sandwich-size bags, for which we charged $1 to $2, paid in cash by the customers at the stand. We also sold bagged greens wholesale to a local health food store, in which case we invoiced for payment by check at the end of the month. Sometimes one of us would staff our booth, sometimes both, and sometimes only a helper.

For extended periods of time, we continued to do the majority of our harvesting in our back yard, in the orchard, and in the small area surrounding the orchard. And because we would go back again and again to the same areas, we began to make some interesting observations.

We saw that each sow thistle plant that we had drastically pruned for salad greens continued to grow and generate new sets of young leaves when other plants were already dying off. (Remember, we never uprooted any plants.) Some sow thistle plants (an annual) produced up to five "crops" before they began to die back in June. Thus, it was apparent that pruning prolongs the productivity of these wild crops for months.

It was also obvious that more rain results in more food. But equally important, we observed that our intimate, intensive, and regular interaction with the wild food plants increased their fruitfulness many times over. Where we plucked off the chickweed tops, at least two sprouts would arise from the cut, and the plant would continue to grow. Where we pinched the tender mallow tops and leaves, the plant would send out more leaves in that spot. Some lamb's-quarters plants could be topped four, five, even six times and would continue to grow and produce abundantly. We experienced this effect with all the wild plants we were picking.

One day in May, we both felt sad as we entered the orchard. We could see that the season was coming to a close. It was getting harder and taking longer to fill a tub with chickweed, and we had to walk farther to get lamb's-quarters, radish flowers, and other greens. After all, the rainfall was becoming infrequent, and we did little or no watering or irrigation, instead just carefully harvesting as the season proceeded.

But the feeling of sadness was quickly transformed when we took an exploratory hike to another local area where we sometimes gathered wild greens. This spot, as well as most of the other wild areas nearby, was already completely dry and lacking in any fresh greens. Clearly, the productive season was already nearly two months past for most of the wild areas in our neighborhood. We are certain that one of the results of our careful, steady gathering was that we effectively extended the season of our wild greens by a couple of months.

We made another discovery. Though prickly pear cacti grow throughout the United States, they are most common in Southern California, the Southwest, and Mexico. In our area, we have huge stands of prickly pear in the mountain canyons as well as along the freeway on-ramps. We are fortunate to have several good patches along the borders of our yard, and we use the peeled pads nearly year-round in our breakfasts and salads. Generally, nearly all of the prickly pear sprout forth new pads in April. The young pads are glossy green compared with the dull, grayish green of older pads, and the spines on the young pads are immature and easy to remove with a dull knife. The first crop is the ideal stage to eat, so every April you see lugs of the cactus pads at all the produce markets (a lug is the shallow box used by sellers of produce; it represents a specific quantity, like a bushel or peck). By June, it's hard to find good cactus in the markets, because most wild-gatherers pick only the first crop.

But we discovered that a second crop grew up in one to two weeks, usually right at the spots where we'd picked new pads before. So we had a good second crop right away. Our patches were so thick and productive that we got all we needed from a small area. Because we were picking cactus for markets at least twice a week in addition to eating cactus every day for breakfast, we visited the patch often and observed the new crops sprouting up within days of the previous crop being picked. It was like magic! We brought fresh, young cactus pads to every market we attended, even when they were absent everywhere else.

Christopher once took a bus from Los Angeles to Cuernavaca, which is about an hour's ride south of Mexico City. The bus drove through some poor villages with adobe houses, where nearly every house had a prickly pear "tree" outside. Apparently, the villagers allowed one or two central trunks to grow up tall, and they pulled off all the younger pads as they sprouted, or the pads were eaten by goats.

At our place we never achieved abundant year-round cactus production for market. By late August and into September, we had to walk through all of the patches of cactus nearby in order to fill our tubs with young pads. Through the autumn and winter we could find enough edible pads for our personal needs—that is, for our breakfast—but not enough for selling. Even so, because we carefully and regularly pruned, there was still food available to us long after the unpruned wild cacti had grown tougher. Though we sometimes pick a young cactus pad, clean it of spines, and eat it raw, we mostly prepare the cactus by cooking. Once cleaned of the spines, we sauté diced cactus pads in a skillet. Slow cooking reduces much of the water from the pad. Then we add potatoes or onions or eggs to make a vegetable dish. We have also added the diced cactus pads to stews, soups, and green salads.

Overall, we became very "connected" to each and every plant we gathered, whether it was chickweed or wild radish or prickly pear. We were saddened as they faded, much as we might feel if a close friend was departing for an overseas trip for the next six to nine months.

We had learned some of the secrets of how Native Americans were able to produce the volume of food they needed. We observed how our frequent harvesting meant more food, not less. And where others saw a field of unruly growth, even when other fields had died back or were plowed, we saw acres of delicious abundance. These wild "weeds" were all so fresh and nutritious that we cringed every time we saw television commercials for home-and-garden herbicides, promising to kill off the very plants we are eating and collecting every day!

By contrast, listen to our sales spiel for the farmers' markets:

"These greens are fresh, picked this morning. Many of them are more nutritious than regular produce. They have never been fertilized, waxed, nor treated with pesticides, herbicides, or fungicides. They've not been genetically engineered. We wash our hands before we pick them and then use tongs or gloves for any subsequent handling. And your dollars don't support greedy agribiz."

We doubt if we would have learned so much had we not done this plant-gathering ourselves. Now we never look at wild areas, nor the weedier areas of our yard, and see only what is there at the moment. We know that untilled but mulched areas will produce abundantly, even though the untrained eye sees no prospect of food. We feel joy every time we walk through our wild areas, an inner joy and sense of interacting with nature's great secrets. By contrast, when we visit the hard-packed rows of exposed soil baking in the sun at a tractor-tilled farm, we don't experience joy. We come away with the feeling that something is profoundly wrong with the ways our society produces food today.

## WILD PICKLES

It's an early morning in June, and we're sitting in our back yard. We lay old carpets out and use them for doing our outdoor exercises. Now we're sitting outside on this unseasonably cool morning having breakfast of toast and an avocado, just listening to the birds and breathing in the fresh oxygen. It is cool and refreshing here, almost like being at one of our favorite spots in the local mountains.

Out back, we spot a small sparrow doing something on top of a tall plant. The plant is sow thistle, a tall relative of dandelion with its yellow blossoms and cottony seed tufts. At first, we think the sparrow is picking out seeds for food—and perhaps she is. But she (or he) is doing something else, picking tufts of the cotton and taking them to a nearby nest in our grapefruit tree. We smile, as we recognize that this is one of the many reasons we allow certain wild plants (what some folks call "trashy weeds") to flourish on our property. The sow thistle is healthy and tall, and the birds seem to enjoy it. Our potbellied pig, Otis, really likes the sow thistle as well, and so does Blue Girl, our goose. And we've both gained a new appreciation for the sow thistle ever since Dolores began to make pickles from the young flower buds.

Because food storage and preservation is almost a lost art, many of the traditional methods of pickling are long forgotten. Refrigeration and supermarkets have made it easy to rely on commercially prepared foods. Having learned a few pickling methods that we can easily do for ourselves, we know that there will always be a place for the skills of "living off the land," whether in a rural or urban setting.

In our neighborhood here in the suburbs of Los Angeles, the open fields fill up every spring with sow thistle, mallow, and wild radish, all common and ubiquitous "weeds." We're lucky that our part of the city is very hilly, so there are many open lots where the developers haven't (yet) been able to build houses. And because few of these landowners value the weeds on their property, it's easy to get permission to pick them. We have used these plants in our soups and salads for nearly three decades, and have always enjoyed learning new methods of preparation.

We decided to do some research into the old ways of preserving foods. We knew that drying agricultural produce was nearly universal in the past. Also common was pickling, which in many cases is believed to increase the nutritional content of the foods.

During a visit to a close friend's home, he insisted that we all try some of his homemade pickles. To our initial surprise, these were not pickled cucumbers. The aroma when he opened the jar was strong and robust.

We each tried a spoonful of the little pickles, or "capers" as he called them. They were pungent and delicious! He told us that they were the flower buds of sow thistle, not only one of the ordinary weeds in our area but a common wild plant worldwide. We were both excited by the prospect of another food contribution from a plant with which we were already very familiar. Our friend told us that he simply packs the washed sow thistle buds into glass jars, fills the jars with raw apple-cider vinegar, and puts them into his refrigerator. He lets the buds steep in the vinegar for at least a month before he starts eating them. The ones we sampled were more than a year old.

This, of course, is not what is meant by "canning," where refrigeration is not needed. Instead this was just an easy way to create refrigerated pickles. Our friend told us that besides raw cider vinegar, he has also used the leftover liquid from a jar of commercial pickles, from a jar or can of olives, or from pickled jalapeños and other peppers. We were eager to try this very simple pickling method.

Since then, we have mainly used four very common flowers or fruits for our wild pickles or capers. There are no doubt many more buds and pods that could be used as caper substitutes, but we'll tell you how we pickle these and you can experiment on your own.

Like our friend, we've used the young flower buds of the sow thistle. Most people think of dandelions when they see a sow thistle, but dandelions never get as tall as sow thistles. We would go into our wild orchard in the early morning and carefully pick off the buds. After a while, you notice that there are buds that look like immature flowers, and those are the best for pickling. Then there are the yellow blossoms, already flowering, which we don't pick. After the flowers close up again they somewhat resemble the immature bud, but at this stage they are getting old and going to seed. Though we have pickled these older buds by mistake, we prefer the young ones.

Second, the mallow plant, also called cheeseweed, grows worldwide and is a fine source of capers. Mallow is also called "poverty weed" by those unfamiliar with its virtues. The leaves often appear in our salads and soups, and its relative, the mallow of the marshes, was traditionally used as a medicine as long ago as the era of the ancient Egyptians. They boiled the root and whipped it up into a white froth to make a cough medicine. Alas, its modern counterpart is a quintessential junk food—the marshmallow. After the plant flowers, the fruit that forms is round and flat. These are good for pickling if you get them while they are still green, not yet matured into dry seeds.

Third, the wild radish plant is in the mustard family, and the flower form is identical to that of the wild mustards: four petals, four sepals, six stamens, and one pistil. However, the wild radish flower is lavender colored fading to white, and its leaves are tastier and spicier than those of most mustards. One uniqueness of the wild radish is the plump seed-pod that develops after the flowers mature. These pods somewhat resemble jalapeño peppers; we have pickled these, then served them to guests who thought they were some sort of serrano or jalapeño pepper.

Fourth of our favorites, nasturtium is now grown by many garden-ers, as it sprawls all over areas with rich soil and some shade. We have nasturtium growing all over the orchard areas, and nasturtium also grows wild along the West Coast beaches. We pick the new flower buds and the two-lobed fruit. For a tasty pickle you have to get the fruit while it is still green and tender, before it hardens.

The simplest procedure for pickling is to begin by rinsing all the buds, flowers, or fruits. This is easily done by putting all the material to be pickled into a bowl and adding lukewarm water, which will loosen any insects and dirt and allow you to pick out any less than perfect pieces. Rinse the remainder in a colander, and put them into glass jars. Then cover them with raw apple-cider vinegar, close the lid, and put the jar into the refrigerator. That's all there is to it! You can also add raw honey, garlic cloves, cayenne, kelp, and a pinch of sea salt.

Though we eat some of these pickles immediately, they improve as they age. Ideally, you should let them sit for about two months before you start using them.

It feels great to be consuming homemade pickles, a product from common plants that most people not only overlook but actually scorn.

And we find it so enjoyable to relate to our wild plants as delicacies, not merely as weeds, while we pick their buds, fruits, seeds, and pods.

Picture in your mind an early morning. It is still foggy and quiet, and the orchard is alive with various birds and insects along with Rocky the squirrel running through the tops of the very tall eucalyptus trees growing in a line along the east side of the orchard. These aromatic eucalyptus leaves, by the way, make a great medicinal tea for any breathing problem, and the bark can be boiled to make an antiseptic tea for treating wounds.

This wild orchard is on a hillside, and we've dug terraces into the slope. We walk through the terraces, where the trees grow in a haphazard way, not in rows, and the ground has been heavily mulched with years' worth of wood chips, sawdust, grass clippings, leaves, and animal manures. Now mallow, sow thistle, wild radish, nasturtium, and New Zealand spinach are all abundant. With a bag in hand, we gently pick off the young sow thistle buds, then blow the mature seeds so the plant will spread even more. We may pinch off several of the older, upper stalks in an attempt to prolong the life of this annual.

We have noted from past years that our extensive picking results in two, three, or even four "crops" from a plant that ordinarily produces just once. For example, fresh sow thistle leaves sprout from each cutting. Mallow continues to produce new buds, flowers, and seeds where we have pinched it back. We scatter the dried seeds for next season.

The wild radish adds lovely white and lavender hues to our hillside orchard, and we find the young seedpods succulent and hotly delicious. The flavor is similar to a commercially grown radish, though the pod looks like a pepper. We walk through the orchard in the morning, carefully inspecting the radish plants, picking off each pod. As the season progresses, we will prune back the tops of these plants and feed them to Otis the pig and to our chickens. Once we do this pruning, the wild radish will continue to grow if the summer is not too hot and dry, which means that there will be even more late-season pods.

And the nasturtiums add color and flavor to the borders of our garden and orchard. All parts of the nasturtium that are tender are edible, and we use them all. For pickles, we take only the small, closed flower bud or the young seed. Both are spicy hot like the leaves and add a horseradishy tang to whatever foods they are added to. The young seed cluster is particularly noteworthy because it has the shape of a brain— divided into two or three lobes—and is very wrinkly. According to the ancient Doctrine of Signatures, this should mean that the nasturtium seeds are good brain food. (Are there any herbalists or researchers out there who know if this is true?)

Making pickles from the wild produce of your garden is merely one of the ways to harvest the bounty just outside your door. Each time we discover a new use for these so-called nuisance plants, we realize how stupid we've all become in our modern attempts to poison and kill the "weeds."

And when the season has changed, and there are no more sow thistle, mallow, nasturtium, or wild radishes to go pick, we'll still have the bottles of pickles to remind us that we encountered the wild flora thoughtfully on its own ground and were generously rewarded.

# 4 **Faunal Friends**

## Animals in the City

*The greatness of a nation can be judged by the way its animals are treated.*

—GANDHI

It's an early summer morning and we're sitting in our living room, looking through our large picture window into our front courtyard to the
south. This season, we allowed a large lamb's-quarters plant to grow outside the front door. No one could believe how large it got, eventually
reaching a height of 12 feet before it began to droop downward from the
weight of the upper branches and seeds. Lamb's-quarters—a relative of
common spinach and quinoa—has leaves that are edible in salads or
soups, and it produces abundant seed at the end of the season. We nearly
always let lamb's-quarters grow when it comes up because it's so nutritious and we use it in so many dishes. But this tall one outside the door
was a showpiece.

And now, as we sit indoors looking out, it has attracted families of
sparrows who seem delighted at the treat they have found. This is better
than any manufactured bird feeder we could have purchased. With most
of our lamb's-quarters plants, we collect the seed for use in soup and
bread, but with this enormous one, we've enjoyed observing the birds
who have flocked to the plant every morning, eventually cleaning the
plant of all its seed. For weeks we could look out our front window and
watch the sparrows having their breakfast while we had ours.

## BIRD-WATCHING

We have always enjoyed watching the array of birds that visit our little
urban homestead. They come to pick up some of the feed we give to our

chickens, and they are also attracted to the variety of seeding plants that flourish here.

One winter, we were talking with a friend who is a professional gardener in Pasadena. We were telling him about all the birds we see in our yard and surrounding neighborhood.

"Birds?" he exclaimed. "What birds? They all go south for the winter!" Needless to say, we were a bit taken aback by his comment. He truly believed that there were no birds around. It occurred to us that he really may not see birds during his daily garden routine. After all, he's a guy whose gardening is "mow and blow" and who always uses chemicals to combat weeds and bugs. Why would birds come to the yards he has "gardened"?

In fact, the presence (or lack) of birds can tell you a great deal about the state of the local environment. Birds generally avoid sterile environments, because they need insects. And because they feed on insects, their presence is good for your yard and garden. Moreover, though the volume is slight, the droppings of wild birds add nutrients to your soil. Birds are necessary for a diverse, strong ecology.

There are many conventional methods to attract more birds to your yard, including bird feeders, birdhouses, birdbaths, and various seeds that they like. Most pet shops and garden centers have numerous bird-attracting products.

We have two large trees where blue jays live. We don't see them much during the day but they must be there, because if we toss a few peanuts in the back yard they seem to know within minutes, and they swoop down to gobble up this favorite food of theirs.

Out back, we allow a tree tobacco plant to grow wild and tall. Tree tobacco is a common wild plant in our local foothills and deserts. It's not a native, but an introduced plant. Visitors have asked us why we don't cut it down, because tree tobacco is poisonous. Once a fire inspector asked us that very question, and we told him that we let it grow because it attracts hummingbirds who eat both insects and nectar from the tobacco's tubular yellow flowers. The fire inspector seemed dubious, and we think he was convinced that this was yet another novel "excuse" for not cutting down the abundant flora in our extended yard. But right then a little green hummingbird appeared, darting from flower to flower. We could barely believe the timing, and the man was instantly convinced.

We sometimes put out store-bought hummingbird feeders to attract these seemingly miraculous birds, which are a joy to behold. To fuel their high metabolisms, they eat not only nectar but loads of insects.

Timothy Hall, who used to be our neighbor here, experimented with countless homemade hummingbird foods. The commercial brands always seem to be red to attract the birds, who are also drawn to brightly colored flowers. Timothy tried mixtures with beet juice and various sugars until he found a mix that he could recommend, although the beet juice did not remain red when exposed to the sun.

Timothy would go to the pet store and buy an 8-ounce container of commercial "instant nectar" for hummingbirds. He'd mix this with water, per the instructions, and end up with 48 ounces of liquid. Then, he'd mix 3 cups of water with 1 cup of raw sugar from the supermarket. "This raw sugar is often described as 'pure raw cane sugar'—it is brown in color and coarse," he explained. He added this sugar and water mix to the store-bought mixture.

"All I've really done," Hall told us, "is extend the commercial mixture so I get more for my money." Moreover, in all of his experimental mixtures that lacked the commercial powder, the hummingbird liquid would get moldy. Mold never developed as long as he used the right proportion of store-bought powder in his mix.

Hall also warned us not to use honey in hummingbird mixes, because the honey can actually kill hummingbirds.


Here in the hilly Highland Park district of Los Angeles, we have seen small hawks on numerous occasions. When we first moved here, we were standing in the back yard one day and noticed a red-tailed hawk high overhead. Just one—that means he was hunting. But then we saw that there were two. That usually means there is a mating dance going on. Then we saw another—three. We assumed that this was mama and papa and baby. Yet as we watched, we eventually counted seven red-tails sailing overhead for perhaps twenty minutes until they eventually circled out of view. Such a rare sight . . . Later, when we told a professional ornithologist about this incident, he expressed doubt that we could have possibly seen hawks overhead in our urban Los Angeles neighborhood. He tactfully suggested that perhaps we didn't know a hawk from a crow!

We hear woodpeckers every now and then, feeding on insects just under the bark of the tall eucalyptus trees out back. It's hard to spot these small birds, but we recognize their familiar jackhammering noise. In most urban areas, dead or dying trees are immediately taken down, leaving no opportunities for woodpeckers. We are fortunate to have the much wilder land out back, with many trees in all stages of growth and decay. We like the woodpeckers' loud sounds, their ruggedness, and their beauty when we do see them. They are remarkable birds who create beautiful patterns of holes in the bark where they have removed their meals of insects and larvae.

Residents of our neighborhood frequently see and hear owls at night. These nocturnal predators keep down the rodent population, and there is also something very comforting about the presence of owls, as if they are sentries keeping watch from the telephone poles at the top of the hill.

Birds are an integral part of the ecology of the city. It is easy to provide some simple forms of cover and food for them. In return, they will eat thousands of insects, as well as provide you with countless hours of enjoyment. We do what we can to invite birds to visit our little homestead, as their busy activity tells us that this little pocket of urban wilderness is alive and vibrant. By contrast, we are often surprised at the sterility of other yards we visit in the city, because we're convinced that anyone who chooses to do so can welcome wild birds into their yard.

We've also created an oasis of hospitality for other kinds of creatures. Most urban areas have laws against "agricultural usage," including raising certain animals considered to be farm animals. However, these laws are typically enforced only when someone complains, so if you would like to raise animals in the city it behooves you to find out the local laws that might apply and then organize your approach to avoid creating any form of nuisance, including excess noise and odors.

We have had chickens for nearly twenty years. We both raised chickens during childhood, and we've each had them in various urban back yards.

Though some people raise chickens to eat them, we raise them solely for eggs and for the benefits of their manure.

The first of our present population of chickens originally came through the U.S. mail in a vented cardboard box. We'd ordered forty-eight of them, one or two days old, from Welp, Inc., a breeder in Bancroft, Iowa, which we learned about in a *Christian Science Monitor* article. At that time, the members of the nonprofit survival-research organization WTI were interested in the Welp line of chickens, which were said to be the best for small farms and also for "survival situations"—that is, they would provide food in a minimal area if you couldn't rely on a grocery store for provisions. After many letters to get our questions answered, we purchased twenty-four of the Welp 650N (they look like the Rhode Island reds, with brown eggs) and twenty-four of their 542 (these look like white leghorns, with white eggs). Though it seemed like a large number to start with, this was the minimum number we could order to get those particular chicks. So forty-eight it was!

To learn the basics of chicken raising, we've used two books: *Grow It!* by Richard Langer and *The Homesteader's Handbook to Raising Small Livestock* by Jerome Belanger. In fact, we've used these two books to answer most of the specific questions we've had whenever we've had problems with bees, chickens, goats, rabbits, geese, or ducks. We don't know if they are still in print, but we've seen copies at used bookstores.

After our chicks began to grow, we had to rebuild their cage several times to accommodate their growing size. Originally they had only a caged yard, then we added a little house inside the caged area, and ultimately the chicken house was rebuilt to be more solidly enclosed. We could not leave our flock in an open yard for several reasons. In addition to the possibility that the chickens might simply leave, there was the likelihood that they would be picked off by local dogs, coyotes, and other predators (raccoons, opossums, hawks, even people) that wander through here, especially at night.

Our current chicken house is a simple pole-frame structure, built by putting corner posts in concrete footings, adding cross-framing for the roof and two doors, then covering the whole structure with chicken wire or plywood. Its present size is 8 feet tall, 8 feet wide, and 20 feet long. The birds share eight nesting boxes and have horizontal roosts high up off the floor. Cleaning involves occasionally hosing down the enclosure on the inside, raking it out, and adding fresh straw or wood chips.

As spring rolled around that first year, we had so many eggs that we began to sell them to neighbors and friends by circulating a flyer throughout the neighborhood. Ordinarily we kept at least two roosters so that we could hatch some eggs. As the male birds grew up and got louder in the morning, we made periodic trips to a local pet store to sell off some of them. We had found out that the zoning for Los Angeles permits hens if they are a certain distance from any neighbor's building, but roosters require more distance because of their crowing. Because zoning regulations vary greatly from city to city, we recommend that you find out the rules for your particular area.

*Eagle turtle hens.*

We learned by direct experience what we have heard so often since—that the common white leghorn chickens are bred for high egg production but a short life and that they are often nervous and are prone to disease and cannibalism. Commercial Rhode Island reds are only slightly less jumpy. Our Welp chickens were better than most commercial birds and were good producers.

Over two years we gradually reduced our flock of forty-eight to around ten chickens, a more manageable number. It's important to be realistic and humane in your ambitions for animal husbandry, because it's detrimental to have a flock of stressed, bedraggled, and nervous birds.

A few months after we'd begun raising our chickens, an interesting drama started playing out in our neighborhood. One day in our neighborhood, someone discarded two living chickens, later identified as red jungle fowl, the primordial chicken that is supposed to have originated in the area we know today as Vietnam.

These two beautiful chickens—a rooster and a hen—lived for a few months in our neighbor Bill's front yard pyracantha bush, a few houses up the street from our place. After some gardeners had done some work in that yard and disturbed them, the chickens took up residence a few doors closer to us. Eventually they found the wild acreage where the owner maintains a wildlife preserve behind our place and took up residence in that area. The exotic chickens lived there for months. They would roost at night 20 feet up in the towering branches of a California pepper tree.

The rooster was big and mean, with large leg spurs. He would occasionally get into our chicken house and kill a hen, presumably by mating and being so rough. We called him the Ayatollah. The hen we called Henny Penny. They were canny, swift, and intelligent—true survivors.

On two occasions, Henny Penny disappeared for weeks. We would only see the Ayatollah and wondered if the hen had wandered off or been killed by dogs. But she was apparently trying to hatch her eggs. On her second try, she was successful. She appeared with eleven beautiful little "game birds," swarming around her as she moved. She quickly taught them the arts of survival—pecking for food, protection, jumping high from branch to branch. We were amazed at how high the little babies would perch.

The Ayatollah had been separate from Henny Penny for so long during the hatching period that it seemed she no longer needed him, or perhaps he had abandoned her. Oh, he would come around from time to time, and he performed some fatherly duties, but he never seemed to be around when he was needed. He spent much more time attempting to get into the chicken coop in our back yard, then attacking our domesticated leghorn hens. He crowed frequently and loudly.

We caught the Ayatollah on numerous occasions, thinking that we'd sell him to the pet store, but we'd always release him, hoping that he'd go and help raise the family with Henny Penny, though he never did. We discussed providing some sort of shelter and protection to Henny Penny and her brood, but always vacillated between valuing their freedom and their safety.

Then in the middle of the night, we heard screeches and cackles from out back, and we knew that something had attacked the wild chickens. We went to investigate and could only find Henny Penny. She was ruffled, but seemed all right. We felt a great sadness: we thought that all the chicks were dead, but in the morning we saw all eleven chicks together. Henny Penny had taught them well. She had apparently fought off the intruder—possibly a cat—while the chicks fled and hid, remaining completely silent.

We still didn't provide any protection or shelter, not being sure where, how, or even if that was the best way of looking after these relatively wild birds. A few nights later, the next assault took its toll. This time, we could see that Henny Penny had been injured and feathers were missing. She had a large gash on her underside. Only six chicks could be found; the rest were never found, taken away we assumed in some animal's jaws. We felt a real loss.

Henny Penny limped now and had difficulty caring for her babies, who were a few weeks old and already swift. The weather was intermittently wet and wintry. We continued to put food out for the wild birds, and we laid a large piece of plywood over some logs as protection and shelter.

Yet we didn't think that Henny Penny would survive without help. We managed to gather the remaining chicks in a small cage, and we took their mother to a veterinarian. Then we kept her inside where it was warmer, and we gave her eucalyptus-pod tea every day. Her entire underside was wounded, scarred, and green. She continued to eat and drink, and slowly Dolores nursed her back to her usual state of feistiness.

It took Henny Penny about two months to heal. She grew stronger, with Dolores regularly tending to her. The gash on her underside healed over, leaving a large scab, which Dolores treated with eucalyptus-pod tea, raw honey, and aloe vera. From that point, her health picked up, so we eventually let her loose in the yard again with her six remaining chicks.

Meanwhile, the Ayatollah had frequently been coming into our chicken yard, attacking the leghorns and keeping us up at night. Somewhat regretfully, we caught him and sold him to the pet store owner, who liked the handsome rooster so much that he took him home as a personal pet.

Gradually, Henny Penny and her offspring multiplied, all living wild in the back open acre, roosting high up in the branches of the eucalyptus trees. And the babies had babies. We once counted more than forty feral chickens out back, and in the morning the noise was getting louder and louder as the roosters competed with each other and the hens were all clucking. It was quite impressive to hear all that early in the morning, but remember, we live in Los Angeles, not rural Maine. Someone wasn't happy about the noise and called the city's department of animal control, who left a notice saying that they might round up the wild chickens if we did not. We didn't care to have guys in uniforms up in our trees and nosing around our yard, so we began to capture them three or four at a time, and we sold them to the local pet store. At least three of the new Ayatollahs were sold in this way.

Frankly, we had lost track of the many generations of wild chickens that were out back. Each year around Easter, we'd notice that we hadn't seen a particular hen, and we'd wonder if a dog had got her. In fact, many chickens were picked off by dogs. But invariably this "lost" hen would appear in our yard with eight or ten day-old chicks. It was obvious that she felt safe here and that she was showing off her babies to us. This always brought a wide smile to each of us, to see the new family of fluffy one- or two-day-old chicks with their mother coming forth in spite of all the dangers she had faced.

After a few years of this, something interesting happened, involving one of the last of the wild red jungle fowl hens, yet another Henny Penny, and one of our last Welp-line leghorns, a rooster named Hector. Hector was somewhat of a back yard sentry. He would jump over the fence and even guard the driveway where people passed by into the property behind us. He would actually kick and scratch you if he didn't want you near, and he developed quite a reputation.

Hector and Henny Penny mated, and Henny Penny was sitting on her eggs in a little spot along the neighbor's driveway under an ivy bush. We hadn't seen her for a while and just assumed that she was sitting on eggs. Then one day our friend Prudence called to tell us that she thought Henny Penny had been killed by a dog. Sure enough, we went out by the driveway and saw that she'd been attacked in a manner typical of a loose dog. After all, she would have been at ground level, where a dog could have easily sniffed her out. We buried her under a fruit tree. We were sad to see her go.

Then, as an afterthought, we realized there were probably some eggs somewhere. We searched along the driveway, pulling back the ivy to see if there were any hiding places in there. We found ten eggs and put them into our lightbulb-powered incubator, and they all hatched in two days. These were a remarkable breed, a cross of the leghorns and the jungle fowl. We called them our Eagle Turtle chickens. We have had several generations of this blend and have found them to be remarkable. While they do not lay year-round, they seem immune to disease, they do not cannibalize, and they live long. We have Eagle Turtles to this day, and at least one breeder has attempted to continue this unique line.

You can take one of the eggs from our Eagle Turtles and one from the supermarket and put them side by side in the pan. The yolks of the Eagle Turtles are solid and golden yellow, contrasted with the floppy, pastel-yellow yolks from conventional chickens. And the whites of our eggs are solid and chewy, not like the rubbery slop from supermarket eggs. We presume that our eggs are much more nutritious, as well.

We know that some folks just matter-of-factly buy some chickens, make sure they are fed and watered daily, and collect the eggs, without getting sentimental about their birds. On the other hand, it's often been said that a sure way to become a vegetarian is to raise animals and give them names.

Ours have all become our friends. Though we were already more or less vegetarians, we have developed a deep fondness for every life-form with which we have closely interacted. It is now difficult for us to go to the supermarket and see frozen chickens and not imagine that these were living, feathered beings with names and personalities.

## BEEKEEPING

Christopher got his first hive in 1979. At that time, he had been involved in outdoor education and teaching people about wild foods and American Indian skills. He was also beginning to plant fruit trees and to garden more seriously. He realized that keeping bees served many functions. He was able to produce his own high-quality honey and sometimes harvested some pollen as well. He was able to produce beeswax at home, used mostly for candle making. And having the bees helped ensure good fruit pollination.

We've never depended on our bees for our main income, and so we never got into the more complicated and expedient business aspects of beekeeping. We viewed that white hive out in the yard as an education—a mystery—offering insights into an important facet of the natural world.

In the beginning, we read everything we could about how to prevent swarms, how to keep ants out of the hive, ways to increase honey production, the bee population, and so forth. We remember the great joy we had experienced initially: to go out in the yard, usually late at night, and press an ear to the hive. It was wonderful, not just the overall buzzing vibration, but the individual sounds that you could pick out from time to time. You would hear a bee scratching on the wood just beyond the wall, and you could hear high and low pitches that deviated from the main buzz. We might sit there for as long as thirty minutes, just listening, noting the patterns of sound and the differences in pitch and wondering exactly how the bees communicated.

*Home of our honey bees.*

CHRISTOPHER NYERGES

Eventually, we learned about the dance that bees do to tell other bees where they've found a source of food. In *The Seven Mysteries of Life: An Exploration in Science and Philosophy*, Guy Murchie speaks about the "group mind" of bees and describes their method of communication.

> This bee language was pieced together little by little in the 1920s and 1930s by Karl von Frisch, the famous Austrian zoologist who has gradually become so intimate with his subjects that, as he says, he literally senses a hive from the inside and "feels himself a bee." It is a

language used primarily by bee scouts who have discovered a source of nectar too far away to be readily findable by smell or sight and who want to share it with their fellow workers. They tell about the new nectar by doing a very precise dance on the honeycomb inside the hive, which consists of quivering the abdomen while going through a curious figure eight maneuver. . . . The meaning of this dance, so painstakingly translated by von Frisch, is that the middle axis of the figure eight gives the direction of the nectar in relation to the direction of the sun, which (replacing light with gravity for the purpose of the dance) is assumed to be straight up or exactly above the top of the comb. Thus if the dancing bee moves up the comb at an angle 27 degrees to the right of vertical, she is telling her sisters to fly 27 degrees to the right of the sun to find something good to make honey out of and, after they have "read" the message several times and understood it, they actually go out and fly the designated course. Even in cloudy weather a bee can usually see the sun's position through her awareness of ultraviolet and polarized light. As to how far bees are to follow the course, the instruction is imparted by the tempo of the dance, a near destination being indicated by faster dancing and a far one by slower dancing, always in direct ratio. And the kind of nectar is also revealed in the dance, this time by smell since the scout has inevitably picked up a sample of it on her legs and body. Besides that, she hums quietly, and not just to herself, the tone intermittently coming from her wing beats at 250 cycles a second which, with mystic appropriateness, turns out to be the key of B. I said "intermittently" because the bee hums audibly only during the middle or "straight run" portion of her figure eight dance which, being a fixed percentage of the whole dance, has a duration in direct proportion to the distance to the new nectar source and therefore tips off all the bees who hear her as to how far away it is.

What an amazing process!

We can't say that we've ever really noted such a figure eight dance, but we have always enjoyed watching the bees, typically by leaning on the top of the hive and bending over to watch workers coming home to the entrance hole. The bright pollen on their rear legs varies in color and volume, and we try to determine where they are getting their pollen. In our neighborhood, we would see eucalyptus pollen, bottlebrush pollen, and other kinds, as evidenced by the colors.

In addition to enjoying the astonishing mysteries of bees, we wanted hives of our own in order to pollinate our trees and to provide not only honey but several beneficial by-products:

- *Propolis: Collected by bees from bark or buds, and called "bee glue" by beekeepers, this sticky mixture of wax, resin, oil, and pollen is used to seal the insides of the hive. Propolis has been credited with antibiotic and bactericidal properties and may be used to help wounds heal. It is included as an ingredient in some natural toothpastes, or can simply be chewed like gum. Propolis is rich in bioflavinoids. Taking it internally (in liquid or tablet form, purchased from health food stores) is said to boost one's energy level and reduce cholesterol in the blood.*

- *Royal jelly: Secreted by the salivary glands of the worker bees, and sometimes called "bee milk" by beekeepers, this is used to feed and stimulate the development of a queen bee. Royal jelly has been collected by humans for centuries because it is rich in vitamins, minerals, and amino acids and is said to provide relief from arthritis and muscular dystrophy. It is antibacterial, helps to control cholesterol levels, and helps to relieve some allergies.*

- *Pollen: When bees collect flower nectars, they also collect pollen on their legs. If you stand next to a hive, you can see the various colors of pollen on the bees as they enter the hive. Small amounts of pollen turn up in some honey. To get pure pollen, a beekeeper must attach a small collector in the opening to the hive so that granules of pollen will be scraped off the bees' legs as they pass. This takes quite a while, which is why bee pollen is so expensive. Pollen is rich in protein and amino acids. It is used by many and is said to increase one's energy, regulate the bowels, boost the immune system, and diminish allergies.*

Beekeeping has also been traditionally associated with relief from arthritis. According to Dr. Robert Brooks, a research physiologist for the

Arthritis Foundation, bee venom raises the body's cortisone level by 1,000 percent, for reasons not fully understood.

According to an article in the *Los Angeles Times*, beekeeper Robert Hoyer of Limerick, Pennsylvania, who has about half a million bees, says that although he wears a veil and protective suit when working his hives, when the arthritis in his back gets severe, he has his wife, Norma, put bees on his back so that they will sting. "My back still bothers me sometimes, but it's not as bad," he stated. Hoyer acknowledges that the medical community is understandably cautious about his form of treatment, and "it's not for everyone." His wife added, "I don't think I'll ever get used to it."

By raising bees, we were able to be somewhat more self-reliant and thus in another modest way "a part of the solution" to some of the problems facing us all today. There is also the economic factor. We don't have to buy honey if we can get it from our hives, and our honey has always been superior to anything we've ever purchased, bar none. We have also sold jars of honey (several gallons annually) via word of mouth to friends, which has meant being able to generate a small portion of our income from our bee hives.

Once we learned about bees' habits, and acquired the requisite beekeeper's gloves and smoker and other equipment, we began to keep between one to three hives at any given time. Before long we got ourselves listed with the local fire department, and we would get calls in the spring to remove bees from people's houses and yard trees. In the beginning, we did not charge for this service, because we wanted to increase the number of our hives. But later, we learned that we didn't always end up with a hive for these efforts, and we would charge to come and remove a swarm.

Over time, we really began to see bees as fellow living creatures, not just commodities or a means of producing income. In one of the jobs where we were hired to remove a swarm from an attic, we found all the bees clustered deep under the rafters in a spot where they could barely be reached. But we'd promised to try, and as usual we made every effort to get out the bees alive, save as much honeycomb as we could, and clean up the attic. It was hot, difficult work and, as it turned out, the attic was full of asbestos. By the time we'd begun removing bees and comb, it was clear that it would be a difficult job to get the bees out alive. We had no idea where the queen was, and the job was too far away from our home for us to set up a hive nearby outside, in hopes that all the bees would go into it. So we began by removing comb, layer by layer, and smoking the bees out. But insulation kept falling into the honey buckets, and soon there was honey all over and so many bees getting back into the honey and the buckets—it became a big mess. We tried to the best of our abilities to get the attic cleaned out and realized that most of the bees would not survive. We didn't like this at all, being participants in killing them. The people should have called not us but an exterminator, if that's what they wanted. It really bothered us to kill the bees, and we never did that sort of job again.

Sadly, we learned that most people didn't want to make the extra effort to get the bees out alive—all too often people just wanted the bees out of their house in the quickest, cheapest way possible. We wanted to keep bees, not kill bees. From then on, we only did swarm removal jobs where the swarm was in a bush or tree, easily accessible.

If a swarm is easily accessible in a low bush, we set up one super (the "hive" in which bees are kept) nearby on a ladder or bench. Then we cut the branch where the bees are hanging, place it in the super, and put on the lid. Of course, it is not always that simple. We always wear protective gloves and hoods. Because most bees find the super to be an ideal home, they'll stay there. After an hour or so, we remove the super to a more permanent location.

Over the years, we did get a few new colonies this way, and we earned some extra income. But overall, we are mostly interested in beekeeping for the lessons inherent in the activity, and because we want to do our part to keep our local crops, flowers, and fruit trees pollinated. Beekeeping is a great service to your local ecosystem.

Bees today are threatened by many factors, including habitat destruction, pesticide use on farms, diminished populations of wild bees, and the fact that many urban people are panicked about the so-called killer bees and are eager to spray poisons on anything resembling a bee.

We think it's important to look upon your bees lovingly. Provide for their needs, and they'll provide for you. When you work your hives, remember that you're in the home of thousands of living beings and that each life is precious. They are not just "things"; if we view them that way, we won't care whether we kill them or if their habitat is destroyed. If we do care, we should be able to see the bigger picture: ultimately the bees' best interests are our own best interests.

There are many beekeeping organizations, and if you are a beginner it is probably best to attend a few lectures and do some reading before you get a hive. Of course, we too have continued to study and read and ask questions, trying to learn more about these remarkable creatures. We've found much fascinating and useful information in *The Art and Adventure of Beekeeping* (volume 1) and *Mastering the Art of Beekeeping* (volume 2) by Ormond and Harry Aebi. The Aebis explain beekeeping through their personal experiences, and so their books are more interesting to read than an encyclopedic textbook. Another good reference book is *The Hive and the Honey Bee,* which answers just about every beekeeping question you'll ever have. We have purchased bee colonies in supers (the layered boxes of a hive) and then added empty supers as the colony grew. You will need a smoker, a few tools for working the hives, gloves and a hood for protection, and possibly an extractor. Our hives have usually not been on our own property but on the property behind us, where we have set them on a secluded spot in a eucalyptus orchard. Our understanding of the laws in Los Angeles is that beekeeping is technically illegal, but this prohibition is enforced only if a neighbor complains. If you strategically place your beehive's entryway and situate the boxes where they are not readily visible by neighbors, a hive or two is unlikely to attract much attention.

*Our Honey*

The honey that we get from our hives is mostly from eucalyptus blossoms. Out beyond the open portion of our back yard there is more than an acre of open land, with many eucalyptus trees. This honey is very dark; sometimes it looks brown, sometimes red, and usually it is so thick you cannot see through when you hold a jar of it up to the light.

It also has a strong flavor. We have learned that whenever we have hay fever, or any respiratory problems, we take a tablespoon of that honey—either straight or in a warm drink such as tea—and we feel better quickly.

We have been told that our honey is "low grade" because it is dark, that it would not win any prizes in a county fair because the light honeys are regarded as more desirable. Consequently, merchants of the past would always sell their light honey first. They kept the "low grade" for themselves, the dark honey with all the pollen and propolis still mixed in. It turns out they kept the best for themselves.

According to *Folk Medicine: A Vermont Doctor's Guide to Good Health* by D. C. Jarvis, M.D., honey is a source of iron, copper, manganese, silica, chlorine, calcium, potassium, sodium, phosphorus, and magnesium. The amounts of these beneficent minerals in a particular variety of honey are dependent upon the character and quality of the soil where the plants were growing from which the bees collected the nectar and pollen.

We use our honey in coffee, drinks, shakes, cakes, muffins, bread, biscuits, pies, and so on. Honey is also a good source of vitamin C and of the simple sugars dextrose and levulose, which are easy for the body to digest and metabolize. We do not buy refined white sugar, which we consider to be a modern poison. In his chapter titled "The Usefulness of Honey," Dr. Jarvis summarizes nine advantages of honey over refined white sugars.

1. Honey is nonirritating to the lining of the digestive tract.
2. It is easily and rapidly assimilated.
3. It quickly furnishes the demand for energy.
4. It enables athletes and others who expend energy heavily to recuperate rapidly from exertion.
5. It is, of all sugars, handled best by the kidneys.
6. It has a natural and gentle laxative effect.
7. It has sedative value, quieting the body.
8. It is easily obtainable.
9. It is inexpensive.

Be aware that it is never advisable to feed honey to very young children, who are susceptible to an allergic reaction.

Though we do not overdo the use of any sweeteners, it's good to know that you can have sweetened foods that are actually good for you.

We use only natural sweeteners—honey, molasses, date sugar, stevia; we never buy refined white sugar or any of the processed sugar substitutes, which our research indicates are as unhealthy as, if not worse than, white sugar. Dr. Jarvis emphasizes that he considers honey best as a medicinal substance, useful for coughs, arthritis, burns, hay fever, and cramps.

As noted above, another use of honey is for treating wounds. Once while Christopher was bicycling home a couple of miles from Pasadena late at night, his tire caught a rut in the road and twisted sideways, and he fell. Because he was wearing shorts, he scraped a large portion of skin from his leg. The scrape really hurt, and his leg felt stiff. He got up and bicycled home. At that time, he lived at a place with absolutely no yard, so there were no plants, therefore no fresh aloe to put on the wound.

The wound was bad, but he washed it and spread on raw honey— meaning honey that hasn't been either pasteurized or heated for straining; raw honey is usually labeled as such. He didn't dress the wound, just left it open. The honey salve felt good, though it was a bit messy and he had to be careful not to get honey all over. In the morning, he dressed the wound, and it healed well.

This is a good example of how we use our research in an ongoing way to address daily problems.

We keep file folders with clippings and notations and index these as an aid in retrieving information later. In our files, we have a *Los Angeles Times* article from the Science/Food/Health section of the Sunday *Home* magazine referring to several Canadian and British studies showing that honey applied to wounds inhibits the growth of infectious microorganisms. Honey had evidently been used in developing countries for many years as a salve for wounds. Nevertheless, according to the article, modern remedies such as sterile bandages and antibiotics are preferable.

We keep track of such issues and maintain files for recurring reference. In the spring of 1985, subsequent studies revealed more about the medicinal properties of honey. At Israel's Tel Aviv Medical Center and other labs, honey was applied to open wounds on mice. Researchers found that hardly any bacterial growth resulted and that rapid healing ensued. These studies showed honey to be effective for burns, infected wounds, and ulcers. The Tel Aviv study was conducted by treating open wounds on twelve mice with raw honey. Twelve other mice with wounds were not treated and were used for comparison purposes. The wounds were measured regularly, and the wounds on the honey-treated mice healed faster. The researchers concluded that the raw honey, applied topically to wounds, accelerates the healing process because of honey's energy-producing qualities, hygroscopic effects on the wound, and bactericidal properties. Apparently the lack of water in honey inhibits growth of bacteria.

## RABBITS

Though we do not currently keep rabbits, we have done so in the past, and they are often described as the ideal homestead animal because they can be raised in a small space, do not make noise, are really odor-free, and require minimal care.

We have friends in Altadena, California, who raised rabbits, then butchered them for meat for the family's meals. They also tanned the skins and used them to make clothing such as vests and hats. However, because we are mostly vegetarian, we raised rabbits primarily for their fertilizer, which is high in nitrogen and great for gardens. In chapter 5, we'll discuss our collaborative rabbit-and-earthworm composting system in more detail.

As we noted in describing the layout of our yard, the rabbit cage measured approximately four feet by six feet, with a fine wire screen for the "floor." All the urine and droppings went directly into the worm farm/compost pit below, which was a fairly ideal arrangement: the rabbit hutch rarely needed cleaning, and the worm population thrived.

We had many beautiful rabbits for several years, but each would occasionally find a way to get out of the cage. Unfortunately, neighborhood dogs would always find them before we could, and we have not continued to raise rabbits.

## GEESE

Blue Girl is our Canadian snow goose, a large beautiful white bird with sky-blue eyes. She has been with us since 1984, and we consider her a part of the family.

She is very friendly in the spring when it's time to lay her eggs, and she enjoys being hugged by those she knows. During the rest of the year, if you are an unfamiliar person she will nip at you.

She lays maybe two dozen very large eggs each spring. They have a rich, strong flavor, and we use them primarily for pickled eggs or for making cookies and cakes. To pickle our goose eggs, we first hard-boil and peel them. We put them in large glass jars, fill the jars with raw apple cider vinegar, cover, and refrigerate.

*Blue Girl enjoys a swim in her pool.*

Geese, like ducks and swans, are waterbirds. At a thrift store we bought a small wading pool for Blue Girl, and she loves to flap around in that. We have since purchased a larger pool for her, measuring about four by eight feet. When it rains, she stands still and holds her head upward, occasionally taking a drink of the rain but otherwise simply enjoying the soaking.

She lives in the northwest corner of our back yard, and along with spending time in her pool she often sits in an old, unused dove cage. Because there is a driveway along the west side of our property, people occasionally walk by, causing Blue Girl (sometimes) to honk her "alarm," which is a benefit. She also honks along with our chickens, and when we had roosters she'd honk with their crowing, causing some neighbors to call her a "goose-ster."

We have been tempted to take her somewhere else where there is a large lake or pond, and no danger of coyotes, but have never found the right place. We have also considered taking her to a lake with an existing goose population, but there is a chance the other geese would kill her. Wild nature can be like this, with brutal death very likely.

It would be hard to part with Blue Girl after all these years, and so we suspect that she will live her life with us, continuing to do her duty as the back yard "alarm" when unfamiliar people walk by.

Otis is our pet potbellied pig. He is black with coarse hairs, about the size of a large dog, and with a belly that nearly touches the ground.

When people see him, they often ask if we are going to eat him. It is perhaps a natural question, and innocent enough, because people *do* eat pork. But we always regard the question as rather gruesome. As with Blue Girl, we think of Otis as a member of our family. He really is an intelligent guy and loves attention and all the food we can feed him.

We initially got him in 1992 when the potbellied pig "craze" was fading and when people were abandoning these miniature pigs to pounds and rescue operations. Otis's previous owner had had a pet shop, but the family was moving and could not take Otis.

We learned a lot on the night we took him home. First Christopher had to sit on the ground, nose to nose with Otis sniffing back, so he could get accustomed to us and wouldn't run away. After all, we were taking him from his home, his familiar surroundings, and his people.

When Christopher picked Otis up and carried him to our car, we learned how awesome a pig's voice can be. It was ten at night and otherwise quiet. Otis squealed as loud as he could, nonstop, and we were amazed that someone didn't call the police thinking that we were kidnapping or even killing someone.

By this point we were having some doubts. When we got him home, it was late and dark, and our neighborhood was completely calm. We did not want Otis to yell again as we took him into the back yard, so we had to slowly coax him to walk into the back where we'd keep him. We shook some carob pods in front of his nose until he'd walk a few feet, and we kept that up until he finally got out back.

Not knowing exactly what to do, we put him in the old dove cage, an enclosure built of 2 x 4s covered in chicken wire, with a floor space of about five feet by eight feet. We decided to let Otis into the back yard during the day, but we discovered that he liked to eat the toxic castor bean plants that grew all over our yard, so we had to confine him and also get the castor plants out of his area.

We knew very little about taking care of pigs when we got Otis, so we purchased some books specifically about potbellied pigs and went to a few lectures. In the beginning, we had Otis sleep in the kitchen under a table, where we tried to confine him with a plank slid in behind the table legs and held up by the adjacent stove. Whenever he managed to bull his way through such barricades, he would do a thorough reconnaissance of each room he got into. Once he found a large bag of whole carob pods in Christopher's office and had eaten nearly half the bag before we found him. Whenever he came into the house after that, of course he went straight to the office to check for more carob. Quite a memory, when it comes to food!

Eventually, we built up a section of the yard just for Otis, using recycled telephone poles for some of the fence posts. He needed a very beefy fence, since he always managed to knock over any weaker barrier and get into the garden. And after Otis had been in the garden for a while, there was no garden left.

Early on, Otis had an old doghouse for his house, big enough for him to get into and barely big enough to turn around in. Soon he had that doghouse fully packed with straw, so it was completely insulated for the winter. But Otis would turn and turn, and with his strength, one day we went out there and the entire doghouse had simply fallen apart. Now he has a large old wooden box, half of the former chicken coop. He still packs it with alfalfa and hay, but he has more room to move around.

Whenever we hear people say that pigs are more intelligent than dogs, we have to laugh. We suppose that that is probably true, if someone conducted some sort of tests of memory and other mental faculties. But you really can't compare dogs and pigs—it's like comparing apples and pineapples. Pigs do seem to have a good memory and are able to be trained more readily, if appropriately bribed with food. But it's clear that they live primarily for food and pleasure. Otis loves to have the back of his ears rubbed, and he really enjoys a good belly rub. But as for "walking" him, rather than walk a pig, you coax a pig, and you can't actually push or force one to go anywhere. The pig walks you.

A dog, on the other hand, seems to live for his or her master. Dogs are often eager to oblige, whereas pigs don't share a sense of that. Our dogs get noticeably upset when we leave for a while, and they often just want to be around us. Otis, for his part, tolerates us as long as we have food. He can certainly distinguish us from strangers, due to his keen sense of smell. And he will snort and grunt if he doesn't like a stranger, for some reason. But a pig doesn't seem to have that loyalty you see in a dog.

Still, we've grown to love Otis. He generally gets fed only a small cupful of miniature pig food each day, and he also gets various kitchen scraps such as banana peels, old bread, kiwi skins, rotten apples and oranges. It is like having a living compost device, able to quickly process much of our kitchen scraps and yard trimmings and convert all this into a top-quality fertilizer. Some things he just won't eat—such as broccoli, or grapefruits, or the peels of citrus fruits—so instead we put those into our regular compost pit with the worm farm.

We soon realized that Otis's manure could be a gold mine. We clean his yard every week or so, filling a 5-gallon bucket with his droppings, then sprinkle wood shavings or sawdust over his toilet area to deal with the odor.

We put the pig manure in the holes when we plant new trees and use it to fertilize old trees on the property. In some cases, we have been amazed to see, within only a few days, the extensive new growth on a tree we had recently fertilized. We have now seen this happen so often, we no longer doubt that it's the Otis fertilizer that provides the impetus for the tree's revival.

DOLORES LYNN NYERGES

We used to use straw for his bedding and in his yard, but the small twigs and sticks often got stuck in his eyes. We switched to alfalfa hay, or grass we collect from out back, and the problem was solved. Plus, Otis eats the alfalfa straw. He will stay up long into the night grazing on his new alfalfa when we replenish his bedding. In a day or two, his belly has filled out noticeably. We laugh when we see him that way, with his "hay belly."

*Otis helps himself to a wild snack of pigweed (lamb's-quarters).*

In terms of care, we have found that we don't need to do much with Otis except feed him, talk to him, scratch him, and make sure he always has fresh water. The one task that's more complicated with a male pot-bellied pig is maintaining the tusks. Because Otis was neutered before we got him, and because he is penned up, his tusks grow less quickly than those of an unneutered, wild pig. When they do get too big and begin to grow back into his skin like an ingrown toenail, there is no way for them to be worn down or broken off naturally, as would happen in the wild.

When his tusks needed trimming the first time, we made inquiries as to how this can be done. A vet said that he'd come over and do the job for $250. Okay . . . but we asked around among various people about how we might do it ourselves, and a friend who grew up in rural Kentucky suggested that we just cut the tusks off with pliers or the type of wire saw used to sever bones during amputations. She made some suggestions for how to do this, including ways of getting Otis used to the wire saw next to his tusks.

Each day for a week Christopher placed the wire bone saw over Otis's tusk and rubbed it a little bit, and it was clear that Otis did not enjoy this. But the tusk was rubbing into his skin, as all the hair on that part of his face had already rubbed away, so the tusk had to be trimmed. Finally, Christopher went into Otis's yard, put the wire saw over the tusk, and quickly began to saw. Otis yelled, as was his custom, and also pulled in such a way that he seemed to be helping the process, as if he knew that the tusk had to be trimmed. It took all of about twenty seconds to actually saw off the tusk, and we have done the task ourselves many times since, every five or six months.

We have some hilarious stories we could tell about Otis. For instance, when we first got him, Otis got sick. Of course we were worried about him and called a mobile vet, who came over to give Otis a shot. That was quite a spectacle. Otis ran all throughout the yard, and ultimately he jumped into Blue Girl's pool, which was full of water. Somehow, miraculously, the vet was able to hold him there and did manage to give Otis a shot, and he got better.

Otis is such a character. We now understand the essence of "pigness," in actual pigs as well as in the realm of symbols and metaphors. We like Otis so much that since we began publishing our *Talking Leafs* newsletter in 1994, we have put Otis's picture on the cover of each issue and ascribed to him some sort of lofty quotation. We would call these "an Otis quotis," and added a subtitle or slogan for the newsletter, saying,

"Known throughout the West as the Otis Notice." This newsletter tells about our classes and the publications and products we offer, with an interesting feature story or editorial in each issue. The newsletter is not about pigs—we just found that it was a good attention grabber to put Otis there on the cover for everyone to see. (Our Web site store is even called "Otis's General Store"!)

By the way, many cities have zoning laws prohibiting agricultural animals due to the noise and odors. A normal breed of farm pig can grow up to be nearly as big as a cow! But Otis is *not* an "agricultural animal." While the city of Los Angeles does not allow farm animals, L.A. has tolerated potbellied pigs because they are clearly not what the original law was dealing with. The city council once had it on their agenda to vote whether or not to make it officially legal to have a potbelly as a pet, and that was the very week we experienced the outbreak of violence and looting that we now call the Rodney King Riots. Needless to say, in the postriot aftermath, the legality of potbellied pigs in L.A. neighborhoods was a *very* low priority. And now that the popularity of potbellied pigs has waned somewhat (evidently one of those cases of a fad coming and going), it is probably unlikely that the Los Angeles city council will ever bring up this subject again. Even so, if you are considering getting involved in raising animals more unusual than ordinary pets such as cats and dogs, look into the animal-related zoning and health department regulations for your municipality. There may be provisions, or loopholes, that permit raising animals under certain specific circumstances.

## CANINE FRIENDS

There is nothing like a dog! Most everyone who has lived with dogs can attest to the many ways they provide enjoyment, entertainment, and good companionship. We have always had canine friends. In our case, the species we fell in love with was American pit bull terriers. Today we have three pit bulls—Cassius Clay, Lulu, and Ramona. Each is part of our family, and each has a unique role.

In terms of urban homesteading, dogs have several contributions to make. Their manure can undoubtedly be used in some gardening applications, such as fertilizing ornamental flowers. We have done research on dog feces and have found that there is no agreement about its use in food gardens, however. Some say that the relative safety of dog manure as a fertilizer in food gardening depends upon the diet of the dog and the extent of *hot* composting used to process that manure. Others say that certain parasites and toxins persist with dog manure even after hot composting. As with most such issues where there are countless determining factors, we feel that the real answer lies somewhere in between. Nevertheless, we do not use dog manure on any of our food crops, except sometimes at the base of fruit trees.

Another obvious value of having dogs around an urban homestead is for defense. Most intruders will avoid a fenced yard with a barking dog. A burglar doesn't want to get bit and doesn't want the neighbors aroused by the barking of the dog. Our dogs have also chased off opossums and skunks, which sometimes get into the chicken house and eat eggs. But because one of our dogs spends the night inside and the other two are in the front yard—without access to the back, or else they too would kill chickens!—they are not much help in protecting the chicken coop at night. Opossums still get in there from time to time, and one night we ourselves had to chase a huge raccoon out of the chicken coop!

We also used to have a problem with squirrels coming into our courtyard to eat avocados. They would just hang out on a limb and chew until the half-eaten fruit dropped. For a long time we weren't sure how to solve that problem, though we have used noise devices and rattraps with limited success. Now that one of our dogs, Cassius Clay, stays in the courtyard, he does a pretty good job of chasing away the squirrels. He has a bloodcurdling bark, and though he can't actually reach up high where the squirrels go, it seems that they have been staying away.

Our female pit bull, Ramah, was just a pup when Dolores brought her home. Until she died in 1995, Ramah was a part of our entire life. We traveled with her, and she guarded the house when we were away for the day.

When we first moved to our home, we used to leave Ramah in the back yard. Ramah apparently discovered how to climb over the fence, and one night she was just gone. We were upset and felt resigned to the fact that we'd never see her again. We assumed that Ramah would be picked up by the pound, or more likely, some person would keep her as a pet, or worse, use her as a fighter in illegal betting matches. Four nights later both of us couldn't sleep, and at about four that morning we walked out in our front yard for no particular reason. Suddenly, Ramah came running down the street. She didn't seem to know where she was and nearly barked at us in surprise. But she quickly recognized us and was very happy to come in with us. After that, she was a changed dog. She had tasted freedom, but thereafter she seemed to appreciate the advantages of home.

Dolores once took Ramah to the high desert, and there she discovered the joys of no fence and no leash. Ramah ran and ran, and you could actually see her smiling and beaming. She was in dog heaven. We can still picture her running for the pure joy of running, chasing those rabbits that she never caught.

Running was always Ramah's greatest pleasure. When Ramah slept, sometimes she would be dreaming—her legs would race, and her voice would whistle and cry. We were always certain that in her dreams she was chasing rabbits through the tall grass.

The first time we took her up to our local mountains for a hike, she raced up and down the sheer hillsides as if they were horizontal, not vertical. We were amazed. And Ramah really shone as our official Wild Food Outing packdog a few years ago. Dolores first made a pack for her by fastening old bicycle panniers over her back. It was very funny to see Ramah walk into Switzer's camp in the Angeles National Forest. After a mile, the pack had rotated around her little body so the top was the bottom, and hikers passing by would laugh and point at her. She loved the attention.

Later, we purchased a regular dog pack. Ramah was so expressive when we put the pack on her that we knew she was very proud and felt important. Everyone we passed along the trail would say, "Look at that dog with a pack!"

As much as Ramah loved to go on all-day walks and to accompany Christopher's classes, she would usually stay in her bed asleep all the next day.

Ramah took a few long trips with us, and our journey to the Intertribal Ceremonial Powwow of 1990 in Gallup, New Mexico, was especially memorable. We paid a Zuni boy a dollar to hold Ramah while we played a drum and orated a piece of writing by Shining Bear. Apparently, Ramah got excited hearing our voices on the loudspeaker, and she broke loose from the Zuni boy. Right in the middle of our presentation, Ramah ran by us and Christopher grabbed her by the collar. She had run until she'd found us. What a dog!

And then there was the rattlesnake bite in November of 1993, when we were coming home from an overnight trip in the local mountains. Ramah had taken in just a small amount of venom, and Christopher took her to a vet. That seemed to cause her health to deteriorate. From then on, Ramah suffered a series of illnesses, each of which she seemed to recover from, but finally we could tell that Ramah was fading in late May of 1995. We didn't want to admit it, but we knew that she was dying. Perhaps Ramah would have lived longer if we had taken her to a vet at that time. But we were afraid that the vet would cut and poison Ramah and make the situation worse, and maybe even that she would die lonely in a cage in some veterinary office. We wanted Ramah to be with us, and we wanted to be with our dear companion.

Christopher promised Ramah that he'd be with her when she died. It may sound odd to some people, but dog lovers will understand why he wanted to be there for her last moment.

One day near the end, Christopher had taken Ramah out because he reasoned that the walk might do her good. But she really couldn't walk well; she was slow and tired. Christopher, too, had been sick and had no energy, but it was the day he needed to collect wild greens for one of the local markets. So both Ramah and Christopher were dragging along for what was to be Ramah's last walk, until Christopher finally decided to come home.

The pain ended at 9:30 in the morning on June 5, 1995. We said our good-byes, realizing there would be no more walks and no more Wild Food Outings with Ramah. We were very thankful for the time that we had had together. When Ramah actually died, Christopher was holding her head in his hands. She cried out a final farewell as she died. He doesn't like to admit it, but Christopher cried most of the day.

A dear friend of ours visited us that afternoon, and we spent the next seven hours discussing death, canine friends, and Ramah. Our friend had brought along a "Dear Abby" article about a boy who had wanted a priest to help him hold a funeral for his dog. The priest agreed. The boy asked Abby if there were dogs in heaven. It was quite an interesting article and made us realize that we weren't the only ones who had such a close relationship with our dog.

Intellectually, we knew that death is part of the cycle of life. And Ramah had become so close to us that we felt she would live forever in our hearts.

But, here's the part where most of us do something that we never feel quite right about. We do it with our human friends, too. We let "the authorities" "dispose of" the dead body of our loved one. With humans who have died, in most cases someone somewhere does things to the body and then it is buried. (A better choice, really, is cremation.) Because there are not the same legalities involved with dead dogs, we knew that we had the option of burying her nearby.

Christopher carried Ramah up to a spot in our "island orchard." This is a place that's very dear to us, a large wild area surrounded by urban sprawl, where we collect wild foods and conduct agricultural experiments. He had dug a large hole in a spot where he felt that a tree might grow. Christopher carefully buried her with her favorite bedding—his old sleeping bag—and then he planted an avocado tree over her. It all seemed so right, so proper. Dogs are not the same as humans, yet our attachments and feelings can be intense.

The "integral garden" we are creating in the orchard out back (which we'll describe in detail in chapter 5) is a place filled with all manner of life. Here, nature is allowed to do what nature does, and the bodies of dead animals can be used to fertilize a diversity of plants. And as Ramah's memorial tree grows, there will be this spot to go and sit, and remember. We recall an old saying, "In remembrance lies salvation." This has many nuances of meaning, not the least of which is that recalling the best of what was and keeping it alive in our hearts allows us to become the type of complete individual that we all aspire to be. Unfortunately, when we destroy nature—whether the wilderness or the wildness of our own back yards—we often lose the place best suited for this remembering. In our case, our orchard garden is such a place, inspiring good memories and waking true joy within our hearts.

A week after we buried our dog, we had a memorial gathering where we invited some close friends to remember Ramah. One of the participants at this gathering, Charles Feibush, wrote to us about a month afterward. "Ramah's memorial was an unusual occasion for me, never having been to a memorial service other than that which one would consider traditional. In all honesty, there was more of a sense of loss at Ramah's burial than at others that I have attended." Another friend who attended the memorial was Mel Hafenfeld. He wrote to us, "It is my fervent hope that we'll once again meet our loving canine companions on 'the other side.' This is the only way I can see that helps assuage the pain of the tremendous loss and which makes any sense in the universe of souls and love."

As you can see, we have many close friends whose canine-pals are also very much a part of their family.

Christopher has spent many hours sitting near Ramah's grave, thinking over what he did right and what he could have done better as Ramah's companion. This experience has made him realize how much greater the pain must be when people lose a spouse, or a child, or any close family member. Sadly, some people never even mourn sufficiently when human friends and relations die.

Even if you're one of those people who isn't "interested in dogs," let it sink into your being that all life is precious. We live here on this earth a relatively short time, and then we leave our bodies and go our own ways.

In telling the story of Ramah, we have gone full circle, from describing her as a close friend—a member of the family—to acknowledging how her body now fertilizes an orchard tree. Where possible, we choose to let our animals fulfill their whole lives right here, on our little urban homestead. Even in death they play their role, promoting more life. Such "complete cycles" are something that many more people took for granted a hundred and more years ago, when most folks lived in rural environments. Today, by our passive choices, we seem to have lost this understanding.

# 5 Integral Gardening

*There is no wiser course in farming than the path of wholesome soil improvement.*

—Masanobu Fukuoka, *One Straw Revolution*

We like the idea, however mythological, of the Garden of Eden, where all the needs of primordial man and primordial woman were met. There were no neat rows, no hard labor, no irrigation systems, no boxes and bottles of herbicides and Miracle Grow. No hoes, no blowers, no mowers, and no weed whackers. In our vision of Eden, plant, animal, and human lived together in harmony. Foods and herbs were abundant. One needed only to go pick and eat. Mythology or not, it is an ideal picture of living on the earth.

It has been said that this myth arose from the hardships of the hunter-gatherer cultures, whose lives were short and difficult, and who toiled to create agriculture little by little. They dreamed, therefore, of a simpler time before nature had come to be seen as an enemy (an attitude that is all too obviously with us to this very day).

How do today's monocrop "farmers" do their work? Large expanses of land are laid to waste, devoid of any living plant or animal, and then large tractors plant the seeds—and spread the fertilizers—and planes dust down the insecticides; then later, if all goes well, other large tractors harvest the crops. The small family farmer of Norman Rockwell's paintings is virtually extinct. Farming today is big business based on "economic principles." Factors such as nutritional content and flavor are a much lower priority than profit margins and rates of return on investment.

Unfortunately agriculture today, as it is practiced on the large scale, may be the single greatest ecological threat to the planet. These concerns have been explored in such books as the *Acres USA Primer*, *The One Straw Revolution*, and *Diet for a Small Planet*, to name just a few.

We enjoy gardening but dislike all the chemicals and fumes and labor and tools and machinery and complexity that seem to go along with "modern" gardening. We do not subscribe to the idea that it is necessary to kill in order to have life.

Between the two of us, we have a diverse background. Both of us have gardened for decades, mostly with the methods known as organic. Actually, while the term "organic" is good, it often implies more work than we do. In general we prefer "natural gardening" or "integral gardening" to describe our approach.

Dolores has run a commercial, organic-oriented gardening business, and Christopher has led classes showing people how to recognize and enjoy edible "weeds." We like to have a yard that produces food, and we have no desire to kill animals or plants by spraying poisons in order to produce that food.

We arrived at the term "integral gardening" to describe what we do because we are not interested *only* in food and herb production. Yet this approach to gardening is effective at producing food, as well. Integral gardening is a way to provide fruits and vegetables by

- *strategically planting fruit trees in your yard;*
- *creating your own mulch and topsoil;*
- *utilizing wild foods and other plants that belong in your climate and ecological niche;*
- *planting those vegetables that "take care of themselves" and that are perennial;*
- *thinking ahead, in order to locate plantings for optimal sun, shade, shelter from wind, and privacy;*
- *and insisting on a no-compromise policy of not using harmful chemicals, whether for killing bugs or improving the soil.*

Some people would say that our small yard appears unkempt and wild; some would see an expanse full of weeds with many unruly bushes.

But it is wild and unkempt only to the untrained eye. Outside the vegetable gardening areas, there are no strict lines, except the "hedgerows," where we have planted along a property line.

The plants we choose usually do more than one job at a time, making overlapping contributions to the plant community, the local ecosystem, and the planet. There is a great deal of mulch, mostly in the form of wood chips and leaves, which we don't rake up but let decompose in place to add moisture and vitality to the soil.

We remove plants such as crabgrass that are undesirable and aggressive, and we feed them to our animals or compost them. There is much life here, and there are no harmful poisons. Even the plants that neighbors and passersby think are "just weeds" are edible, medicinal, or fragrant, or all of these at once.

How do you convince neighbors in an average urban area that this kind of experimental gardening is somehow better than the usual "mow and blow" method? If you aspire to abandon conventional ways, you need to be respectful, diplomatic, and persuasive. Some neighbors will want to learn, some will take more convincing than others, and some will be too set in their ways to pay any attention. It may be best to keep the most visible parts of your yard "normalish" so as not to offend anyone or cause unpleasant legal difficulties.

And neighbors who are open to the power of a good example will recognize the value of your yard's heightened biodiversity and well-being. For example, compared with even the yards of close neighbors, our yard contains many more birds. Some of these birds come and feed on the leftovers from grain we give to our goose and chickens, but many more are attracted by abundant seeds from the wild plants or come to graze on aphids or other insect pests. Integral gardening recognizes that the health of the wild birds and other critters is just as important as your own health and well-being.

Another contrast with other places in the neighborhood is that our yard is cool and shady in the summer, not hot and barren. There is a certain peacefulness about the yard that is unique and hard to convey. Visitors are often surprised that a little haven like this can exist in the city of Los Angeles. One man even suggested that he might take a vacation and camp out in our back yard.

Integral gardening is not a *no*-work system. In fact, in the beginning, you may find it to be more work than you anticipated. You will need to be thoughtful and watchful about what you do. You will need to be alert to the weather and to subtle changes in your yard and among your plants. You may need to undo the previous residents' mistakes. Vigilance is necessary so that you learn to recognize signs that indicate a need to water, to mulch, to clear away particular plants that are out of balance with their surroundings.

Integral gardening is a responsible, responsive way of thinking about your own little plot of land—your designated "small urban farm," where you safely recycle water, create oxygen with your trees and other plantings, and in your own way realize the Garden of Eden dream (sans the beguiling serpent, we hope).

Over the years, we have seen or heard of many different "systems" and "techniques" for gardening. Many are labor intensive (for instance, French-intensive "double digging"), and some are very costly. Some of the backyard gardens you see in slick magazines cost thousands of dollars!

We have also heard of various gurus and experts, including several self-proclaimed "fathers of organic gardening," who seem to want to claim that they are the first ever to have practiced an extraordinary method of gardening, though often they are advocating nothing more than a particular slant on some ancient technique.

Over the years we have tried some of the experts' systems, or parts of them. While we're not against any of these, per se, we have found that, for us, narrowly systematic approaches to "gardening" are limiting and counterproductive. Some of the advocates of gardening systems have beautiful plots and produce to show for it, and we are always impressed when we see huge beans and handsome tomatoes. But sometimes, the celebrity gardeners are more interested in self-puffery and personal fame than in the actual joys and challenges of gardening. And all too often, it appears that these folks spend much more time at their garden-related tasks than we care to spend.

We know from experience that we can simply plant a tomato in good soil, and the plant will grow and produce delicious organic tomatoes.

French double trenches and other labor-intensive methods may be great for the seminar circuit and wonderful if you have a team of volunteers working off jail time. But we quickly tire of all that pointless digging. We just constantly improve our soil and make thoughtful selections of plants to grow. We also recognize the fact that, as with wild nature, some survive and some do not. Our integral garden is home to survivors, those plants that survive well in our area.

When we teach integral gardening classes at the local college, we start with the basics. Frequently students come in and want to know why their corn doesn't do well, or why their tomatoes get so full of bugs. We try to avoid questions that require much specific knowledge of all the details of the yard and garden in question, focusing instead on major underlying principles that can be applied anywhere.

For example, if you wish to have healthy, hardy, insect-resistant, nutritious, flavorful, frost- and drought-tolerant plants, you need to concentrate on improving the soil. Plants get most of their needs from the soil. If the soil is poor, your plants will never thrive. *If you do not grasp this fundamental truth, you will forever miss the essence of integral gardening and will always be preoccupied with side issues.*

There are several ways to begin at the beginning and build the soil. These approaches include compost, homemade fertilizers, manures, mulches, natural pest controls, and companion planting.

Let's look at these one by one.


COMPOST

Ever hear that old saying about a glass being half full or half empty? We cannot count how many people have told us that the soil where they live was no good for gardening when they moved in. We have the same story. Those who see the glass as half empty don't bother to learn what the soil needs, how to improve the soil, and how to adapt to their local conditions of light, wind, temperature, and seasonal change. They try more watering; they try insecticides; they try artificial fertilizers. But what is actually needed is to improve the basic health of the soil.

Those who see the glass as half full have the same soil problems as everyone else, but they have the positive outlook to do something about these problems—all of which can be solved.

In terms of flora, our place was truly a wasteland when we first moved here. That is, as we have said in previous chapters, this place *appeared* to be the "badlands" of the neighborhood. The front yard was like hard-packed adobe, where the previous residents had parked cars, and the back was a neglected barrens covered in tall, dried-up grasses. But because we had had so much practice converting poor soil into rich soil over the years, we knew what needed to be done.

One of our first tasks after arriving was to build a large compost pit. Our back yard is divided into upper and lower sections, the lower being the level where the house sits. Up on the higher level, just beyond the drip line of a grand old grapefruit tree, we laid four railroad ties in the ground as the perimeter of a rectangle measuring about five feet by five feet.

The daily routine is very simple and would work just as well in an apartment with a rooftop or balcony garden as in a house with a small to medium-size urban lot. In the kitchen, we keep a large can with a lid in which to put kitchen scraps. When it is nearly full, we pour it into the compost pit.

In our case, we made the perimeter of the compost pit large, because we wanted it to serve two functions. We needed it big enough to receive our daily kitchen scraps (we separate those bound directly for the compost bin from those that will first be fed to our animals for "precycling"). We also wanted to raise earthworms. You can't accomplish both of these goals in a small space. In small compost piles with the right moisture and mix of ingredients, the temperature will rise so high that any earthworms that don't flee would be cooked to death. But in our large pit, we are able to add fresh kitchen scraps and yard trimmings to one area, and the earthworms just naturally stay on the fringes.

Once our pit was dug and the enclosure installed, we added a small bucket of redworms *(Lubricus rubellus)* to the slightly disturbed, bare soil. Although there are bigger earthworms, these redworms are used by growers because they proliferate the most rapidly while tolerating the broadest range of heat and cold and the greatest variety of wetness and dryness. Earthworms constantly burrow through the soil, breaking down vegetable matter into smaller bits. Their burrowing aerates the earth, and their "castings," or droppings, are rich in nitrogen, both of which are essential to healthy soil.

Soon after moving in, we were given a few rabbits by friends, so we built a large hutch over half of the compost pit. The hutch's four legs were 4 x 4-inch posts that rested on the railroad ties that enclosed the compost pit. The bottom of the rabbit hutch, which was about two feet off the ground, was composed of rigid screening with a small mesh. This meant that the rabbit droppings and urine would go immediately into the compost pit, and we rarely had to clean the rabbit hutch. Because there were earthworms working the soil in the pit, there was also rarely any of the odor you often get around rabbit cages. It was the ideal overall situation, with the rabbits "feeding" the worms, and with top-quality soil produced in the compost pit in only a few weeks.

We have no doubt removed hundreds of gallons of wormy compost over the years, which we've used when we've repotted plants and replanted various trees, shrubs, and vines. In the process, we have gradually transplanted earthworms all over the grounds, which meant that the poor soil we started with has little by little gotten better. Adding compost in quantities of about 10 percent by volume to your garden soil is generally all that is needed to increase the health and insect repellency of your trees, vegetables, and other plants. It isn't the compost itself that keeps insects away, but the fact that compost supplies plants with the nutrients they need, because it's been proved that insects eat "weak" (vitamin- and mineral-deficient) plants.

In case the rabbit and earthworm system sounds complicated, be assured that there are many ways to create your own compost using raw materials you are discarding daily: food scraps and spoiled leftovers from the kitchen, and clippings and refuse from yard maintenance.

Composting is the epitome of simplicity. Organic matter will compost over time no matter what you do, so there's no need to fret that you aren't composting correctly. With a little extra effort—checking to be sure the pile is moist and aerated enough to facilitate bacterial activity and adding brown (*carbon-rich:* dry leaves, rotted hay, small sticks or plant stems) and green (*nitrogen-rich:* grass clippings, garden trimmings) in a C:N ratio of 20 or 25 to 1— you can create a higher-quality compost faster.

Two books from Rodale Press, *The Complete Book of Composting* and *The Encyclopedia of Organic Gardening*, are both excellent sources for determining the nitrogen, phosphorus, and potash (NPK) amounts of commonly used animal manure fertilizers. Another excellent data source is *A Homesite Power Unit: Methane Generator*, by Les Auerbach. This book is a how-to book on building a methane generator, and there are some good tables on the carbon/nitrogen ratio of various manures and compost materials, as well as an NPK analysis.

We've observed some gardeners who take their kitchen scraps and grass clippings and merely turn them into the soil here and there with a shovel or pitchfork. This will work, of course, but it means that you will have vegetable matter in different stages of decomposition throughout your yard. This might mean that your yard will attract flies, mosquitoes, hornets, mice, and an assortment of other undesirable creatures, seen and unseen.

We recommend making fully composted soil first, in a specified area, then distributing it through your yard.

We have tried numerous composting techniques over the years, and there are a wide variety of containers that you can make or buy. For example, we have taken an old plastic trashcan, cut out the bottom, and fastened an old window screen securely in the bottom. Then we set the trashcan in the yard where it was convenient, and added kitchen scraps into the top. The lid fit securely and was always kept on. Because the trashcan was black, it absorbed heat, and decomposition was always quick. In fact, many of the compost makers that are now commercially available are nothing more than complicated versions of our simple trashcan with a screened (or no) bottom.

When we've had large amounts of yard trimmings and grass clippings, we've constructed larger composting devices. One of these consisted of four wooden pallets forming a square enclosure; we nailed three of them securely together, and then we attached the fourth side with hooks and eyelets so we could open it. This we filled with crabgrass, tree prunings, and old cornstalks, vines, and leaves. We covered the top with an old rug, which we occasionally watered. This worked best when the material we added was already broken down into smaller pieces; you can chop up yard wastes with a square-bladed shovel. On occasion, we have used a chipper to chop up larger branches before composting them. It may be worthwhile to rent a heavy-duty chipper if you have a big yard with many prunings and fallen sticks and branches to dispose of. Remember that you don't actually have to compost wood chips—you can simply spread them on the surface as a mulch.

Another variation was to nail four sides together, each about the size of a pallet, but instead of wooden slats the sides were square wooden frames over which we tacked chicken wire. We used this composter for large volumes of yard trimmings, and we found that it worked slightly better than the four pallets, presumably thanks to better aeration.

Within the past ten years or so, we've noticed that we get charged a "disposal fee" for old tires whenever we purchase new tires for our vehicles. We have gotten into the habit of having the mechanic put the old tires in the back of the car and we don't pay the disposal fee. We have used these old tires in various ways—as steps in our back yard, going from one level to another, and as steps in our orchard garden. Once they are placed where you want them, and filled with soil, they seem to last and last. We have also used them stacked flat in several layers as a retaining wall.

A visitor once inquired whether or not these tires would pollute our yard and garden, because she had heard that tires cause air pollution. The tires that contribute to air pollution are those that you are driving on—once tires are worn down, where do you think all that rubber went?—and of course those that are burned as refuse. Tires in a garden decompose very, very slowly, and we're sure that a dozen or so tires in the yard represent no hazard at all. We have found tires to be better than cinder blocks or rocks for steps in most circumstances, because they provide a broader tread and you won't hurt yourself if you slip. Tires can also be adjusted to various positions with ease, and once tamped with earth they are extremely stable.

We have also experimented with using a stack of tires for a compost pit. We reasoned that the tires would stack easily, would absorb heat, and would hopefully create compost quickly. After trying this many times, we've concluded that stacked tires are better as growing pots for certain plants, for instance, tomatoes or potatoes, and that while stacked tires will decompose raw materials, they don't permit air circulation, so other composting systems are easier and more efficient overall. Although everything seemed to decompose well within the inner hole of the tire, the material that got into the tire itself often did not decompose well. That, coupled with the fact that it was often difficult to get all the material out of the tire, moved us away from tires as an easy, streamlined way to make compost. You might experiment with tires—because they are so plentiful—and see what results you get in your area.

In addition, we have used the sheet-metal outer shell of a discarded water heater for a composter. This was a piece of intriguing junk we picked up off the street, and it seemed to work adequately. However, we found it just a bit too narrow for our purposes, and too tall to load and turn easily, although someone might find that rolling the long cylinder around the yard might be a way to turn the contents sufficiently.

*The Complete Book of Composting*, mentioned above (originally published in 1960 and periodically updated—the edition we have is a full 1,007 pages thick!), offers many more suggestions for making compost, including composting in plastic bags, composting within an approximately three-foot-diameter roll of chicken wire or woven wire fencing, and composting within an area enclosed on three sides with stacked cinder blocks, bricks, or fieldstones. The book provides illustrations of many designs, ranging from casual to quite stylish (for instance, a four-sided, picket-fence-style compost pile), and includes many arrangements with two or more compartments, which allows you to fill one area, then let it decompose while you fill the second and in some cases a third. By then you are ready to use the decomposed material in the first bin. Legendary homesteader Scott Nearing had a system of *seven* piles in continuous rotation!

Some composting systems use a cover, while some do not. A top prevents access by pests, such as rodents, and helps keep the pile moist, though with a covered pile you may need to add water, depending on your climate.

In 1968 at his parents' home in Pasadena, California, Christopher dug a small hole to start a little compost pile in the space between their garage and the neighbor's garage. For decades following, the family would add kitchen scraps, leaves, and grass clippings to the hole, which was gradually enlarged to a perimeter of about two feet by four feet, several feet deep. A heavy carpet was always kept over the top of the compost pit, held in place by a few rocks. This kept mice and flies away but allowed in rainwater. To add material, a rock would be removed and the corner of the carpet lifted up. This simple compost pile (which is still there to this day) proved to be all that was needed to process the regular organic wastes from that household. Fresh materials were added at one end and periodically taken out as finished compost from the other end; this compost was then sifted into buckets and scattered around the yard, under fruit trees and ornamentals.

Frankly, there are so many methods of composting that there's no reason why everyone shouldn't have a compost pile or bin in the yard. Most households end up with lots of leaves, branches, so-called weeds, lawn clippings, and kitchen scraps, all of which can go into the composter instead of into landfills or trash incinerators, especially in this day and age of "tipping fees" for household trash disposal, thereby saving money, helping prevent pollution, and avoiding loss of valuable nutrients.

We find it shameful, if not downright idiotic, that people wash away so many useful and valuable resources via the "garbage" disposal, and likewise hire "gardeners" to mow and blow their lawn and leaves, then toss those useful and valuable resources into the trashcan. The way we all deal with such resources is truly a reflection of our spiritual values and condition. Meanwhile landfills across the country are filling up, in areas that were once pristine forests, canyons, and valleys. Thus, there are far greater rewards to recycling than merely being "ecologically cool," or "saving a few bucks."

## HOMEMADE FERTILIZERS

Some people have told us that they do not garden because it is "an expensive pastime." That outlook seems backward to us. At former times nearly all people gardened because homegrown vegetables, fruits, and even grains were far better and cheaper than those purchased from a store.

In the era before World War II, when chemicals came into widespread use, every farmer—whether of large acreage or an urban lot— knew that to produce healthy plants, you had to improve the soil. If the soil is weak, your plants will be weak and subject to insect infestation.

There are many low-cost methods for making your own fertilizer. One of the easiest and best is to decompose seaweed in fresh water.

We learned a great deal about the beneficial properties of seaweed from Ernest Hogeboom, who used to be a professional gardener in the Pasadena area. Hogeboom would collect several plastic trash bags of kelp from areas along the Pacific Coast. He would then empty the kelp into a 55-gallon drum, fill it with water, and cover it. As the seaweed began to decompose, the water would turn brown. Within about two months, the plant fibers were fully decomposed. This liquid was used as a concentrate, which Hogeboom would dilute with water before spraying it on or pouring it around his customers' plants.

Dolores used this for our landscaping and garden-service clients, with the addition of fish emulsion (approximately a quarter cup of fish emulsion to a gallon of seaweed elixir. Plants sprayed with this mixture seem to repel insects and generally show some renewed growth. The only pitfall is the fishy, oceanic odor that is detectable for a day or two after the application.

Seaweed is rich in potassium, up to 12 percent by volume. Though seaweed contains many beneficial trace elements, it is relatively poor in nitrogen and phosphate, which is why the addition of fish emulsion makes a nearly perfect fertilizer.

Rather than use the bulky 55-gallon drum that Hogeboom used, we purchased a 30-gallon plastic trashcan at a building supply store for under $10. This has served us quite well.

While in coastal areas seaweed is readily available for only the expense of hauling, if you have to buy it, seaweed can be costly. Yet if you live far from the seashore and are in the habit of buying all sorts of liquid fertilizers and other commercial treatments for your garden, you will be happy to learn that at least two commonly discarded kitchen scraps are ideal for many of your garden plants.

You've heard of liming the garden and lawn, right? Many gardeners buy a bag of lime (calcium carbonate) every few years and sprinkle it throughout the garden. Were you aware that eggshells are 93 percent calcium carbonate?

Calcium is an essential plant nutrient that plays a fundamental part in cell manufacture and growth. Most roots must have some calcium at the growing tips. Plant growth removes large quantities of calcium from the soil, and so calcium must be replenished. In addition to calcium, eggshells contain about 1 percent nitrogen, about 0.5 percent phosphoric acid, and other trace elements that make them a practical fertilizer.

We save our eggshells in a pan in our oven when we aren't baking in it, including shells from the eggs we buy at the store and eggs from our own chickens and from Blue Girl, our goose. The pilot light temperature of the oven is sufficient to slowly dry out the shells. Then we crush them by hand, or powder them in the blender. Then we place the crushed eggshells around fruit trees and roses, or in potted plants, and broadcast them throughout the vegetable garden.

CHRISTOPHER NYERGES

Snail problems can be solved with the help of recycled eggshells. Instead of powdering the shells, use them at the hand-crushed stage, with plenty of rough, sharp edges. Scatter the crushed shells in circles

*Dolores waters an open-pollinated heirloom tomato.*

around those plants that the snails are eating. Because the shells cause discomfort to the snails, they nearly always retreat and will not cross this jagged protective barrier. (By the way, did you know that California brown snails are actually small, escaped *escargot*? They were brought here to be raised for food and got out, and are now ubiquitous. One method of control is to eat them—but that's another story.)

Another commonly discarded form of kitchen refuse is coffee grounds. Used coffee grounds contain about 2 percent nitrogen, about a third of a percent of phosphoric acid, and varying amounts of potash, generally less than 1 percent. Analysis of coffee grounds shows that they contain many minerals, including trace minerals, carbohydrates, sugars, some vitamins, and some caffeine. They are particularly useful on those plants for which you would apply an "acid food," such as blueberries, evergreens, azaleas, roses, camellias, avocados, and certain fruit trees.

We dry our coffee grounds in the oven, too. Then we scatter them lightly, as a mulch, around those plants that we feel would most benefit. We don't scatter them thickly when it is wet outside, because coffee grounds have a tendency to get moldy.

Because most plants need calcium for root growth, most can be beneficially stimulated by adding a mixture of lime (ground-up eggshells) and dried coffee grounds.

Smile the next time you drink your morning cup of coffee and eat your breakfast of eggs, because the by-products of your meal are ideal for an urban garden and need no longer be seen as kitchen "waste."

## MANURES

In general, for someone in the city with limited time and space, and not a lot of extra driving-around time after work, the best fertilizer is your compost pit and whatever else is easily and readily available. On the other hand, if you can find inexpensive sources of animal manures, they provide nutrient-rich additives for gardens and landscaping plants.

When it comes to animal fertilizers, one of the best, relatively easy-to-come-by manures is rabbit droppings. Rabbit droppings have the highest percentage of nitrogen of any of the commonly available barnyard manures; rabbit manure is about 4 percent nitrogen, compared with 0.5 percent for cow manure, 0.7 percent for horse, 1.8 percent for chicken, and 0.3 percent for hog (depending on which source you read, these percentages can vary considerably). Rabbit droppings are small, compact, and nearly odorless. One organic gardener described them as "miniature, time-release fertilizer capsules." If you raise rabbits, or know someone who does, you'll have a steady source of one of nature's richest natural fertilizers.

As explained above, in the past we have had our rabbit friends living atop the earthworm compost pit. Rabbit droppings can also be called "earthworm caviar"!

Many old-timers in the northeastern United States have been using a recipe for liquid "manure tea" as a fertilizer for years. Variations of this recipe have been around for a very long time. A key factor for the home gardener is the low cost and availability of the materials, especially if you live anywhere near a farming area.

Use any available barnyard manure—horse, cow, rabbit, poultry, hog, or sheep—as long as it is at least two weeks old and therefore not "hot" enough (in terms of actual heat and also the presence of ammonia) to burn plants. Put a shovelful in a 5-gallon container, fill it with water, and let the mix stand for at least a week, stirring occasionally. Then strain the solution through a piece of insect window screening or cheesecloth into another container. Because the dark brown liquid may be too concentrated to use directly, dilute it with enough additional water to make it a light brown color—similar to that of tea.

Christopher's uncle Thomas Jonke lived on the family farm in Chardon, Ohio. Outside his barn, he kept a 55-gallon drum, which he would fill about a third full with composted cow manure. He would then add water and keep the drum covered. When he wanted to water and fertilize his tomatoes, corn, and other garden plants, he would dip into the barrel with a large, #10 can that had a clothes hanger for a handle. He'd then pour the liquid onto those plants he wanted to fertilize. He would continue to add water and use the brown solution for fertilizer. Eventually, all the manure would get used up, and then he'd dump another bag of composted cow manure into the bucket and start over. That was Uncle Tom's manure tea (may he rest in peace).

We are often asked about the relative value of dog feces as a fertilizer. After all, if you are an average urban dweller, you will have access to a steady and abundant supply. We have spent much time researching this subject on the Internet and through calls to vets and others and have collected a great deal of conflicting data and opinion.

In general, the quality of the diet determines the quality of the feces as fertilizers. This is as true for dogs as it is for all other animals, including humans.

There is considerable debate as to whether or not a vegetarian diet has an effect on the safety of dog feces. Those who argue that vegetarianism is no advantage are assuming that a dog is fed commercial dried dog food from bags, and they seem to believe that no pathogens (nor the secondary hazards introduced during manufacture) survive the dog-food-making process. But people also feed their dogs table scraps, including meat and old eggs, and some of these could be a source of problems.

There are known toxins that you can expect to find in dog feces, and although the heat from compost pits may be sufficient to destroy some of these, that may not always be the case. Remember, no two dogs are fed the same foods, and so all conclusions are nothing more than estimations, not reliable facts. And while extensive studies have been conducted on farm manures, dog manures have apparently not warranted that level of consideration.

In our case, we choose to act with caution—using a variety of animal manures but taking care with how we manage the decomposition and where we utilize the results. Because we have three dogs, we do have a steady supply of dog feces to deal with. When we collect it, we have used some in a hillside "fill" area, where we're reestablishing vegetation to inhibit erosion but where we won't be collecting plants for food in the conceivable future.

We would not use dog feces fertilizer on root crops or greens, but it does seem okay to use when planting trees, especially ornamental trees and shrubs. We have buried a bucket or so of dog feces in a hole where a new tree would be planted. We reasoned that it would be maybe four years before we'd be eating the fruit from that tree, and that this should be sufficient time for any toxins to break down. We have also added dog feces to a dedicated composter and then, months later, used that compost for fertilizing trees, again these being plants from which we will not be eating. This, in our case, has included fragrant ornamentals, bamboo, willow, kapok tree, some roses, eucalyptus trees, oak trees, and others.

When it comes to the subject of "humanures" (a term popularized by Joe Jenkins, author of the comprehensive guide *The Humanure Handbook*), again diet is a key factor. In many of the countries where "night soil" is commonly used as a fertilizer, the diet is rather different from the average modern or Westernized diet. We would go so far as to say that if you eat a normal American diet with no regard to additives or the purity of your food, and if you ingest various legal and illegal drugs, and if you drink alcohol, you should regard your feces as *toxic* and should never consider using them as compost.

## MULCHES

A mulch is a material that you lay on the surface of your lawn, garden, or orchard in order to hold in moisture and hamper the growth of unwanted plants. A mulch can be organic or natural substances such as grass clippings, straw, wood chips, pebbles, rocks, or shredded leaves, or can be a manufactured material such as plastic sheeting or recycled carpet scraps. Some mulches (such as wood chips) absorb water, and some (such as gravel) do not. A mulch is not something that should be composted, or that will attract rodents or bugs. For example, you would not "mulch" with banana peels or apple pulp.

Although compared with adding compost or trucked-in topsoil, mulching alone will take a longer time to improve the soil. We have found that mulching with layers of ordinary grass clippings is an easy way to improve the soil and increase the number of earthworms. Layers of grass clippings are quite effective as well as being free for the labor of mowing and raking. A good, thick grass mulch will usually be dry on the top and moist where it contacts the soil. As the clippings decompose, the upper surface of the soil will retain moisture and become host to all sorts of insects and earthworms.

In our experience, a thick mulch of grass clippings will be ideal when you are improving the soil's basic quality, starting from an overall poor state of health. As the soil improves, you should reduce the thickness of the mulch, eventually using thinly spread grass clippings or old straw scattered over the surface of the soil, in order to keep the proliferation of harmful insects to a minimum.

Back before you could recycle telephone books, Geraldine Hogeboom told us how she and her husband, Ernest, used those telephone books for mulch. She would tear out about twenty-five pages at a time, covers and all, and lay them on either side of their vegetable rows in the garden. Then they covered the sheaves of pages with a light layer of soil and wet it down. Geraldine said that this worked wonders in keeping down unwanted growth, and generally the pages would be totally decomposed in two to four weeks. They would observe that the soil remained more moist under the mulch, which no doubt aided in attracting earthworms.

As explained in chapter 2, when we were attempting to restore our long-abused, nutrient-poor grounds to healthy soil, we had private tree-pruning companies dump many truckloads of wood chips in our yard, or in an adjacent yard, and then distributed them throughout the lawn, orchard, and garden areas. In the garden, we are very selective about what sort of chips we bring in. For example, if a load is mostly eucalyptus, even if there's no charge for the material we will turn it down, because too much eucalyptus oil will retard the growth of vegetables.

We have fed our ducks by maintaining a thick layer of wood chips and grass clippings in their area. They were happy as could be, digging for grubs and earwigs and pill bugs that had taken up residence in the mulch.

We have used hundreds of gallons of sawdust in a garden area whose soil needed more dramatic improvement. Aged sawdust is better in an active garden, because there is a temporary loss of nitrogen available to surrounding plants as sawdust decomposes. In our garden we are often thinking long-term, and in areas where we're not actively producing food that season, we have spread great quantities of sawdust. Not only has this helped the soil maintain moisture, but it has led to the growth of many mushrooms during the wet season. Several of these wild mushrooms we have eaten, including *Agaricus campestris* and inky caps. *Obviously, you should eat no wild mushrooms unless you have studied mycology and can positively identify the variety.*

Our wood shavings and sawdust come from the wood shops of two friends. Every month or so they call and tell us their barrels are full, and we go get the sawdust and scatter it where needed. They like the fact that they don't need to dump a potentially valuable resource, and we like what their sawdust has done to our gardens. Needless to say, we do not use sawdust from pressure-treated wood in a garden or as a mulch, since such treatment involves toxic chemicals.

PEST CONTROL

We cannot emphasize this principle enough—to improve the soil is to improve the quality of the plant life living on and in that soil. Part of the plan of the universe is that bugs eat not just any plants, but *weak* plants.

You may or may not be able to tell by looking whether or not a plant is weak—lacking something it needs in the soil or in its environment or maybe having too much of something.

When you are overly concerned about bugs and merely attack them, you are making the same mistake made by a person who gets sick and runs to the doctor, crying "Doctor, please, cure me!"—then who takes a load of pills, perhaps ameliorating the *symptoms*, but in no fundamental way altering the conditions that led to sickness.

Bugs on your plants are nothing but a symptom. Your plants need something! That is what any insect infestation is telling you.

That said, we acknowledge that it may take several years to build up your soil to the point where the plants grown there are naturally insect-resistant. In the meantime, you may wish to eat. Here are some safe techniques we have tried for combating insect problems.

For many years we have grown certain herbs (mints, oregano, arugula, and others) and members of the onion family (including garlic) all over the yard. Herbs and plants such as onion and garlic are alleged to be not only insect-resistant in their own defense, but also to actually repel insects. Truth be told, it seems to us that sometimes they do, and sometimes they don't.

But while we certainly have insects in our gardens, they are never a serious problem. By growing a variety of plants—not in rows but in patches—we have found that we don't get major infestations of insects that may occur when you have all one crop, as is typical of monoculture farming. Diversity (interplanting of different species) discourages excessive concentrations of insects in any one area. Other related strategies include crop rotations and companion planting, which we discuss below.

We have experimented with natural sprays, primarily solutions of garlic. You can grind up garlic or onions, add water, and blend them in an electric blender, sift the solids, and spray this on plants with an insect problem. We have found that this works on most plants, most of the time, to repel insects, and because this mixture is also a safe fertilizer, there are no harmful results.

We have often read about the benefits of making an insect spray by collecting a problematic insect and grinding it up with water in your blender. This, too, Christopher has tried, and though he has noticed some positive effects, the results were not pronounced enough for us to continue the practice, especially considering the unpleasant task of collecting and grinding up insects.

In the past, we made our own insecticide by cooking wild tobacco leaves in a pot of water. We would use about three cups of leaves to about two gallons of water. Once the water turned dark, we'd strain out the leaves, add about a teaspoon of liquid dishwashing detergent, and use this as a direct spray for insects. The nicotine in the tobacco makes the spray effective. This is one of the reasons that we allow a huge tree tobacco plant (Nicotiana glauca) to grow in our back yard. The other reason is that it attracts hummingbirds nearly year-long.

But we have not used that tobacco pest deterrent in years, because we haven't felt any need for sprays. We certainly do not advocate the use of any "universal" insecticide, meaning some substance that you just spray on to kill any and all insects. While unfortunately all too common, that approach demonstrates ignorance and insensitivity. Instead, to really address the causes of a problem, you need to observe the specific characteristics of an insect infestation occurring in your yard and then take steps to deal with that specific phenomenon. There are very comprehensive books that assist the home gardener in identifying specific garden insects and help you determine what action to take.

Sometimes the best response is very simple, costs nothing, and directly modifies the pests' behavior. Picking bugs off plants and crushing them is an example. On occasion, we have observed many aphids on our roses or large amounts of scale on our grapefruit. On warm days, when the garden hose was laying out in the sun so the water was warm, we have simply sprayed off the bugs with direct spray from the hose. This has often been all the insect control we have needed.

As another example, we have had very good results using diatomaceous earth to control chinch bugs in lawn areas and to control snails and slugs around sensitive plants. This powdery white substance works by drying out the snails, slugs, or other insects—its tiny, jagged mineral particles puncture the insects' bodies. (By the way, we have also used diatomaceous earth directly on our dogs. We rub the powder into their coats, which dries out the fleas and gives the dogs a period of relief.) This substance is mined. It is not biodegradable, because it has already decomposed as far as it can. We sprinkle the diatomaceous earth around the plants we wish to protect and are careful not to raise it as dust, as it can be very irritating to breathe. It is very effective for short periods of time, generally not more than a month, and even less if the weather is very wet.

We have used crushed eggshells in the same manner as diatomaceous earth for slug and snail control. If you eat eggs, and have only a small garden, this would be an ideal use for the shells. When mixed in with the rest of the compost, they take a long time to break down unless you crush and grind them first.

In recent years, many gardeners and farmers have also experimented with using "biological controls"—introducing or enticing a predator to aid in the battle against pests. These helpers are a more subtle and perhaps gradual response to an infestation than the drastic reaction of chemical pesticides.

In the early 1990s, Mediterranean fruit flies (a.k.a. medflies) were found in our area, as well as in various other places throughout southern California, and this caused state officials to fear that the agricultural business (specifically the citrus industry) might be threatened. Ignoring vehement protests from citizens, the state adopted the severe action of spraying residential areas with malathion.

On numerous evenings, whether you liked it or not, helicopters flew overhead releasing this sticky red substance intended to kill the medflies. No one really knew what we were getting sprayed with, because we were told that the spray contained about "90 percent inerts"—remember, those so-called inerts are sometimes more lethal than the malathion. The government reasoned that medflies would be attracted to the bait being released by helicopters and would be killed by the malathion. According to many well-informed scientists, this method of medfly control had very little chance of really killing medflies, but everyone who was paying attention noticed a dramatic decrease in bee and ladybug populations following the sprayings. We were both very active in trying to legally stop the malathion being sprayed over residential areas. We prepared newsletters and educational handouts, attempting to teach people about the life cycle of the medfly. It seemed to us that the aerial-spraying campaign was more the result of money and politics than a sound scientific rationale.

And while the use of malathion in Los Angeles and surrounding communities may have eased fears about the citrus farmers' export crop in a psychological way, this was by no means a good thing for the local ecology.

Once the spraying was over, we began to release ladybugs in our yard and back area. We did this several times, because ladybugs are one of those "good insects" that consume "bad insects." We informed friends and students about the use of ladybugs and encouraged people to release many of them in our area.

Other insects that are worth releasing in your yard for eating the "bad insects" include the praying mantis and the lacewing fly, both of which are now commonly available at garden supply stores and in catalogs. When you purchase one of these insect helpers, you'll get instructions on how to release or set it out in your yard.

But just releasing insects on a one-time basis is not the best solution to the problem of poor plants resulting from poor soil. Insects come and go, and they won't necessarily stay in your yard; once the pest problem is under control, the predator insects will move on, seeking food.

We keep emphasizing this point: the solution to all your garden problems is to enrich the soil constantly.

Think of it this way—humans are no different. If we make the analogy that fertilizers for the soil are akin to education for our minds, then we recognize that we should not be content to "go to school" for a few years and then forget about further education for the rest of our lives. We should constantly be learning, striving to improve our state of mind and state of body. Only by so doing will the fruit that we humans produce be of the highest quality, whether that "fruit" be our words, actions, or creations.

## COMPANION PLANTING

Plants are like people in many ways. One way is that they like to be around other plants, with very few exceptions. We have found that the garden that is "happiest" overall is one where there are no rows, and where the plants grow thickly together, providing shade for each other, as well as the diversity that results in greater protection against insect infestations.

Many gardeners have observed over the years that certain plants seem to thrive together and that the proximity of specific plants provides a synergistic benefit to all the plants in the "community."

For example, there is the traditional Native American trinity of plants—the "three sisters" as they are called in many regions: corn, beans, and squash. In very different climates, Hopi and Iroquois farmers, for instance, found that these plants support each other in a mutually beneficial manner. They would first plant the corn and let the stalk grow up. Then the squash would be planted, and its sprawl on the ground created a moisture-conserving mulch of broad leaves. Then the beans would be planted next to each corn plant, so that the beans could use the cornstalks as a trellis for climbing and the nitrogen-fixing nodules at the roots of the beans would provide fertilizer that the corn needed. This way of thinking reflects a system that can involve other traditional food plants as well.

The "three sisters garden" is just one example of complementary plant companions. Another set of plants that like to be near one another is onions and potatoes. Onions seem to keep bugs away from potatoes. Actually, any member of the garlic and onion family can help deter undesirable insects. We recall a funny expression to help you remember the relationship of onions and potatoes: plant onions near your potatoes so the onions will make the eyes of the potatoes water. It's just a silly joke, of course, but the underlying point is that these two do well together.

In general, as noted above, members of the squash family (including cucumbers) do well as ground cover, acting like a mulch and a moisture-trapping canopy for the soil.

You can plant herbs throughout the garden. They tend to be insect-resistant, they break up the monocultures, and other plants will benefit by being near them.

Our approach to companion planting is quite simple. We know from experience that single plants staggered over an otherwise bare soil do not prosper. Plants do best growing closely together with many neighbors, as they would in the wild, with fallen leaves and blossoms for natural mulch and fertilizers, all supporting each other and trapping moisture that all can use whether the summer brings extremes of wet or dry weather.

Perhaps the best treatment of this subject is in Rodale's *Encyclopedia of Organic Gardening*, which provides descriptions of a large number of plants that are known to be healthier and more productive when grown together.

## HEALTH AND THE GARDEN

Let's now step back a bit from scrutinizing specific gardening techniques in order to take a deep breath and consider the many benefits to your whole being that can arise from working with the land. For some people, this involves being open to an entirely new way of thinking.

When people attend our Integral Gardening classes, we tell them right away that if they're only interested in the question of how to grow tomatoes with the least effort, then they're in the wrong class. There are hundreds of books and dozens of TV shows and videos that tell you the nuts and bolts of how to create a garden, organically or not.

Just as some biologists have begun to regard our entire planet as an organism (this is known as the Gaia principle), we suggest that you regard your garden, your yard, your block, your whole neighborhood as a living entity. We are more accustomed to thinking of rural or wilderness areas in these terms, but this perspective is just as applicable to urban communities. If you were taking a biology class, you might go to a wild area and discuss the ecology of the local birds, rodents such as squirrels and mice, fish, and larger mammals such as coyotes; their relationships with the plants, geology, water patterns, and weather. You would begin to see the wider cycles of nature at work. The death of any living thing feeds others, and every species and each individual seems to have its place in the web of life. Deciduous trees lose their leaves, which cover the ground and decay, providing food for herbaceous plants and mushrooms. Small animals feed on grasses and roots; then they are eaten by larger predators, who in turn are eaten by larger predators. And so it goes.

Likewise, think of your yard as an organism. Let the dropping leaves

## Herbs and Vegetable Companions

The Sycamore Farms in Paso Robles, California (2485 Highway 46 West, Paso Robles, CA 93446; toll-free phone number 800 576-5288), publish a newsletter called *Sycamore Gazette*. In their April 2000 issue, they discuss some of the relationships between certain herbs and garden crops. The following is excerpted with permission:

- BASIL: Plant with tomatoes to improve growth and flavor, and to repel flies and mosquitoes.
- BORAGE: Good companion for tomatoes, squash, and strawberries; deters tomato worms.
- GARLIC: Grow near roses and vegetable plants to repel aphids.
- HYSSOP: Good companion for grapes and for cabbage; deters the cabbage moth.
- PEPPERMINT: Good companion for cabbage, beans, carrots, and sage; deters carrot flies, beetles, and cabbage moths.
- SUMMER SAVORY: Said to improve the health of onions and beans; discourages cabbage moths.
- TANSY: Plant with fruit trees, roses, and raspberries; deters Japanese beetles, striped cucumber beetles, squash bugs, and ants.

from the trees feed the tree itself and whatever else lives around it. Let the miracle unfold and take place before you.

We feel that it is important to recognize that human willfulness tends to be a disruptive and counterproductive force. We all know people who want their yard to look a certain way, according to some notion of "beauty," probably based on some picture they saw in a magazine, with no regard for what is truly right for that location—that *specific* location. We urge our students to work with what they already have—what would be growing in their yard naturally—then seek to improve the diversity and health of the plant community by improving the soil. Whatever you do, don't just go in and cut everything down, like so many of the "developers."

We remember a particularly poignant moment in the movie *Out of Africa*. Meryl Streep's character was asked why she didn't just drive the "squatters"—the native Africans—off her farm property. She responded that she would never do that, because "they live there." Our perceptions of the meaning of land "ownership" and property "rights" have become so distorted that we somehow believe that ownership confers any and all rights upon us. Meryl Streep's character was not so hardened to think that way; she knew that the ancestors of some of the Africans had lived on that land, just as generations before had done, long before she was "owner."

We have often been asked why we don't cut down the poisonous castor bean plants that are so common in our yard. Though we do try to control their growth somewhat, we have come to realize that "they live there" in our yard. We don't know why other neighbors don't have castor beans in their yards, but this plant has been here way before we arrived, and now we've learned to accept it. Castor is a tropical plant that grows quickly and tall and provides generous amounts of shade. The seeds, though poisonous to consume, are beautiful. We have taken the time to explore many of the medicinal values of processed castor beans—castor oil—and have also looked into the plant's symbolism, which has long been known as *palma christi*, or "palm of Christ." This is an interesting name. We know the "palm" part is because the castor has a palmate (handlike) leaf, but we are not sure where the association with Christ came from.

We keep these ideas in mind when we see the castors all over our yard. If we were merely allowing our human will to direct what we grow and don't grow in our yard, we're certain we'd have eradicated this plant. That is what most "developers" do, and that is even what most home-owners do, to their local flora. By wiping out those plants that "live there" because of some criteria of relative desirability or relative "beauty," we all lose, and we deprive ourselves of potential lessons.

In your own gardening and landscaping, make the most thoughtful plan that you are able. Then follow that plan for all plantings, gardening, and improvements, but watch the results, and seek to learn from nature. Adjust your plan, changing it whenever you see that a different approach would be better for promoting diversity, soil and plant health, and animal and bird habitat.

As emphasized throughout this chapter, when it comes to creating a wonderful habitat for humans and others, small and subtle measures are often more beneficial than drastic actions. We practice recycling all the time in the garden and yard, not simply as "a way to get rid of trash" but as a way of minimizing what we remove from our yard, because whatever is taken away is lost resources. Anything that can be decomposed is processed in this way to be returned to the soil, and anything that can be fed to our animals is given to them; then their wastes are returned to the soil.

We have a large bucket filled with rusty cans and water. We scoop out the "iron water" as needed and give it to our plants. (This is more extensively described in chapter 8 in the discussion of metal recycling.)

Also, we have probably used just about every imaginable object for a planter at one time or another. For temporary planters, we have used large and small cans, plastic and cardboard milk cartons, hubcaps, buckets with holes, old crates, coconut halves, tires, bags, and even shoes. In time, many of these decay to soil (even the metal cans), or they just decay and break apart, in which case we put what's left into the city's recycling bin.

When we pot larger plants such as trees, big bushes, or bamboo into big 15-gallon containers or larger, we put some cloth or fabric into the very bottom of the pot in order to hold moisture. This gives the plants a bit of an edge during the heat of the summer. Guess what happens to a lot of our old clothes?

Beyond the many "techniques" we've learned, we try to remember our principles—more difficult to describe, perhaps, but confirmed and reinforced by ongoing experience.

An essential principle to consider is that loving, conscientious work puts Light into the soil, the garden, the location. When you work in accord with the principles of the universe, and you work lovingly, your own inner Light is increased, and you thereby affect your environment in a wide range of positive ways. Everyone and everything around you can be elevated by good work in the right spirit.

Though most urban people consider themselves to be not "superstitious" and may not believe in such phenomena as nature spirits and devas and elementals, there are people who have interacted with these beings for centuries. We believe that they are there, although urban sprawl with its concomitant zeal to make every bit of land organized and under control makes it less and less likely that nature spirits are present. But it is possible to create harmonious pockets of wildness where these entities can reside. It is they who often give a certain special "magic" to a site.

There are specific practices that we have used when working with the garden to encourage the presence of nature spirits. For example, we have practiced certain intonations, or singing, while sitting in the garden area. We have noted on occasion that the floral beings actually sway in time to our incantations—"believe it or not."

We have also done thrice watering, where we water all the plants three times, not just once. There are a few reasons for this, not the least of which is that the first watering does not soak into the soil, but only gets the surfaces wet. It often takes that third watering for the roots really to get any benefit from the watering. And while so doing, we contemplate the meaning of threeness, the power of the triune.

We also interact with our plants through a breath exchange. Do you remember what plants "exhale"? Their exhalation is the oxygen we breath, and our carbon dioxide exhalation is what they need in order to thrive.

Practices like these three examples are good to do; we believe that they are as important to gardening as good composting, mulching, and companion planting. The key is to do them in a posture of Lovingness— which is hard work! But not as hard as struggling with a garden that's being forced to submit to human will.

One of the main premises of integral gardening is to "nourish the soil." In our experience the soil can be nourished by spiritual means. This should not mean that you just "think" or "pray" for good soil and abandon all the useful physical efforts and horticultural strategies. The spiritual and the physical need to be working together, in balance, cooperatively.

As a wise one once said, "The physical, without the spiritual, is empty. The spiritual, without the physical, is effete."

# 6 Water

*The Sea is not merely around us. It is inside us. For life on Earth, and perhaps life elsewhere, is primarily organized water. Thales intuitively understood this in the fifth century B.C. when he voiced his extraordinary surmise that "water is the one essential element of the world."*

—GUY MURCHIE, *THE SEVEN MYSTERIES OF LIFE*

Water is one of the most basic essentials for life, in all situations and all civilizations. Water renders a location habitable, whether that be a weekend camping site or the place where a vast city arises. A look into the past shows that rarely did large urban centers develop if there was no ready source of water, for example, the great Egyptian civilization that developed along the Nile and that depended on annual flooding for agriculture. What became Los Angeles was originally the Yangna Indian village, set alongside what is now called the Los Angeles River. The Indian village grew into a Spanish town and eventually into a major city.

As the population of this sprawling urban center increased, water had to be obtained, and thus you have one of the most complex stories of this region—how water was brought to the people. If you have ever seen the movie *Chinatown* with Jack Nicholson, you've seen a fictionalized account of how lands and water rights were bought up so that Los Angelenos would have water.

Thus, if you live in Los Angeles, and likewise in many other metropolitan centers, your water is piped in from far away. In the case of Los Angeles, at least 87 percent of the water comes from the Colorado River or California's Central Valley. It should be sobering for Los Angeles residents to realize that they live in a *desert*, with almost continual scarcity of water. For those of us living here, it is therefore a matter of survival to get to know local water sources and learn how to find water, how to store rain, how to purify water, and how to reduce water usage.

Just like Los Angelenos, most city dwellers are only faintly aware of the origins of their water, that precious elixir of life. Urban areas are usually supplied by sources such as rivers or lakes, reservoirs, or underground aquifers, which may be distant. In some areas, residents are blessed with indigenous springs or seasonal mountain runoff from local hills, but most are not so fortunate. Although purchased springwater is expensive (more costly per quart than gasoline!), many people, urban and rural, are now becoming accustomed to that expense. And the great drought of the late 1980s in the northeastern United States—which now seems to be reoccurring in most parts of the country—has shown that all of us, not just the residents of the relatively drier western states, need to be conservative in our use of water.

A number of filtering techniques developed for campers can be used to ameliorate some of the unappetizing (and hazardous) characteristics of municipal water. Various distilling devices (including home-scale solar versions) offer some benefits near oceans, where the possibility exists to desalinate seawater. And perhaps the most abundant and most underestimated water source available to all city dwellers is rain.

We will first survey the purification and filtration options, then consider several ways that you can collect and store rainwater for a variety of household uses, including as an emergency source of potable water. Then we'll look at water conservation measures that everyone can incorporate into daily routines.


## WATER PURIFICATION

Water experts tell us that any open sources of water anywhere in the United States should be treated before drinking. That is not to say there is no pure water, but with surface water the likelihood of encountering impure water is high. This includes all lake or stream water and even springwater unless you know the source of the spring.

You cannot look at water and determine that it is safe to drink. Beautiful, crystal-clear, quick-flowing water may be full of biological contaminants that will make you sick. And, ironically, a desert waterhole that appears stagnant and covered with insects may contain water that is safe to drink.

Typical causes of water contamination, even in wild areas, include activities of people—swimming and operation of motorboats, dumping of refuse, washing out diapers, and washing dishes. Even surface water that is not used heavily by people may be unsafe, due to airborne pollutants, contamination of the groundwater source, or naturally occurring biological hazards associated with runoff from pastured farm animals and wildlife (fecal material and dead animals).

If you ever suspect your water of being contaminated, boiling is the easiest and cheapest way to purify water. Any hobo can pick up an old can and use it to boil water. According to U.S. Forest Service hydrologist Mike McCorison, who is familiar with the water situation in the California forests, boiling kills every living organism that can make us sick. Of course, boiling does not deal with chemical contaminants, merely biological contaminants. Distillation is the only method of purification that purifies water from chemical as well as biological contamination.

## Distillation and Purification

Distillation is the best way to purify water, because only pure water distills out. However, distillation setups are generally not portable, nor cheap.

If you already have a kitchen distillation unit, that's good! These devices are sold at appliance stores and are most often used for distilling tap water. However, because most of these run with electricity, a high-tech distiller may be of no use when you need it most urgently. We have seen some low-tech distillers that need only a flame to operate and some that actually work with concentrated light from the sun. We haven't used one of these ourselves, but they are worth investigating.

There are also chemical methods of water purification, such as the pills and tablets you can get at backpacking supply stores, with new products becoming available all the time in this field. We don't claim to know all the products on the market, but we would argue in general that none of the water-purification pills on the market are advisable. Though some of these may be effective when fresh, most people don't use up a whole bottle of water purification pills at one time; instead they save the pills for future use. Once opened and exposed to heat and air, these pills are usually not efficacious beyond a year or two.

Our research indicates that iodine crystals are the cheapest, most efficient, and longest-lasting chemical means of purifying water. While some pills and liquids have a shelf life of up to two years, iodine crystals are effective in a broad variety of water conditions, and they remain viable indefinitely. Iodine-purified water tastes better than chlorinated water but does take some getting used to. However, some people should not use iodine crystals, including pregnant women and anyone who has a thyroid condition, because iodine concentrates in the thyroid.

To purify water using this method, add about 4 grams of USP-grade iodine crystals to a 1-ounce glass jar with a plastic lid (Bakelite is best, because iodine water will not degrade this type of plastic). Fill this small glass jar with water and hold it in your hands for a few minutes so that the water remains at body temperature. The water inside the jar will turn a golden color from the iodine crystals. Then, measure out 10 cc of this concentrated iodine liquid and pour that into a quart or liter of water. The water in the bigger container should be safe to drink in about five minutes. We use a lid on our small glass jar that holds a volume of exactly 10 cc, so the lid itself can serve as our measuring device. Keep in mind that you must be careful to pour out 10 cc of the gold-colored liquid and never add the crystals directly to your drinking water, because the crystals themselves are toxic. Once you use up the liquid in the small concentrating glass with the crystals, you simply add more water. You can keep doing this for as long as you have crystals remaining, which could be up to ten years.

In the past these little glass concentrating jars and iodine crystals could be obtained at some chemical supply stores or from camping gear suppliers. For more information, contact Survival Services at the address given in the resources section in the back of this book. Recently, because iodine crystals are used in some types of illegal drug production, the federal government has begun to restrict their availability.

*Filtration*

Another method of water purification utilizes filters.

There are many filtration-type water purifiers on the market, many of them designed for backpacking, and most are very effective. We use a brand called Timberline, which is a pump filter, quite easy to use. It is made primarily of plastic with a cylindrical filter and retails for about $20. There are others that are more durable and cost a little more, for instance, the one called First Need. The best model, in terms of overall sturdiness and effectiveness, is the Katydyne, a device used by most Red Cross personnel when going into areas where the water purity is questionable. This company manufactures several models, and the one most popular with backpackers will cost you somewhere in the neighborhood of $150 to $200 (depending on where you buy it and whether or not it's on sale).

Water purification straws are another option worth considering. As the name implies, these look like straws, and you simply suck the impure water through the straw, which is supposed to remove impurities. These cost less than $10 and are convenient to use. But we have been told that you should regard these as useful only on a temporary basis, for several days, say; then you should discard them. Trying to save money, people may hold on to these filters after they've used them, but they may not be effective anymore.

It's important to acknowledge that sometimes the unexpected happens: you need to purify water, but you have no purifying devices and you can't make a fire. In such situations—whether they are the result of an urban disaster or a wilderness mishap—it's worthwhile knowing about some of the truly makeshift filters that people have devised over the years.

CLOTH FILTER. If you are not seeking potable-quality water and are simply trying to filter out mud and debris, a clean cotton kerchief will work just fine. Pouring muddy water through the cloth will remove most of the unwanted materials. Moreover, you can just let a container of muddy water set for a few hours and the heavier solids will settle out. If you may be tempted to drink the water you acquire in this way, remember that the simplest and the best basic method to purify water is boiling.

SAND AND CHARCOAL FILTER. The 1938 *Boy Scout Handbook* describes an interesting filtration system that can be easily constructed if you have certain basic materials. You need two buckets or barrels (size determined by your needs) connected by a tube or pipe near the bottoms of the containers. Your impure water is poured into the first barrel, which has been filled, from the bottom up, with a perforated plate a few inches up from the bottom (basically, a metal plate or disk with holes in it), a layer of charcoal, a layer of coarse sand, and a layer of gravel. You leave about one-quarter of the space at the top for adding the impure water. The water passes through the layers of materials, then passes through the tube into barrel number two, which is filled, from the bottom up, with a layer of coarse sand (the pipe or tube from the other barrel comes into this coarse sand), charcoal, and gravel. The top third of the second barrel should be left clear for your filtered water.

Such a filter is effective for removing much of the grit and sediment in the water and probably works better than allowing the water to settle. The charcoal may help to remove some pathogens from the water. If you've got on hand the buckets or barrels, tubing, and the materials to put in the barrels and you do not have way to boil water (or cannot have a fire), then this is a low-tech system worth considering.

PEAT-CHARCOAL FILTER. Here is another low-tech water filter, tested in a lab by Stefan Kallman some twenty years ago when he invented this technique. Kallman's design uses an empty beer or soda can or container of similar size, which can be found just about anywhere, whether in the outback or urban jungle.

Remove the lid from the can and make a few small holes in the bottom. Put a layer of sphagnum moss in the bottom of the can—fresh green sphagnum, recently picked, with most of the water pressed out of it (sphagnum can be bought at garden supply stores or foraged in the eastern United States or Europe). Next grind some charcoal until most of it is fine powder, though a few small chunks are all right. Do not use barbecue charcoal that has any lighter fluid in it; you can make your own charcoal by burning hardwoods in a campfire. Mix the charcoal with peat (foraged or purchased in a garden center). You can also mix in some sphagnum. Blend the three ingredients and pack them into the can until you have filled it about two-thirds full. Then add a layer of small pebbles to the top.

One must carefully pack the materials into the can and then carefully pour suspect water through the filter. The first few pours won't be filtered well yet, and you'll note that the water will initially be grayish with particles from the loose carbon. After a few pours to flush the filter, you will be able to drink the water.

Keep in mind that this filtration method was tested with peat, sphagnum, and charcoal. The charcoal is mostly carbon. Carbon has long been used in various water filters to absorb pathogens. The sphagnum chelates, or chemically traps, some metals and also absorbs some pathogens. The layer of pebbles is also important. Without that, water will tend to form channels in the softer materials below and will pass through without significant filtering.

We have seen many variations of this "primitive" filter, and these are worth investigating. Though it might seem that there are easily available substitutes, we'd caution you to not drink water filtered with other materials until you actually test the resulting water. Our advice is to keep a good, reliable, tested water filter at home, and remember that the easiest and cheapest way to purify suspect water is by boiling it.

## RAINWATER COLLECTION

Historically there have been many communities (settled peoples as well as nomads) who have depended to varying degrees on rain for their water supply. Yet many of us in the modern era have lost an awareness of how viable rainwater collection really is. We ourselves have harvested more than 400 gallons from a single storm, without great effort and at virtually no expense. Rainwater is perfectly acceptable for outside use (for instance, landscaping and gardening) as well as washing and even drinking, if filtered and tested.

The technology for capturing rain varies from culture to culture, depending on available materials, patterns of rainfall, geology, and the reliability of other water sources, among other factors. Some forms of architecture incorporate special designs that make it easy to collect and use rainwater, for instance, the Hawaiian catchment roofs described and illustrated in Michael Potts's book *The New Independent Home*.

The simplest rain-collecting device we've seen consisted of a large plastic sheet measuring about four by eight feet, a few clothespins, and a few 5-gallon jugs. After at least a half hour of heavy rain (to allow most of the impurities to be washed out of the air, and most of the detritus off your roof), you're ready to begin collecting water from the sky.

Attach the plastic sheet with strings or twine to bushes so that it is stretched out somewhat, and secure the edges with clothespins to keep the sheet spread out. A sheet extended like this gathers more water simultaneously than would fall directly into the container. Set up the angle of the sheet so that all the water runs down to one point. Place one of your water jugs under this flow of water, and in a short while the jug will be full. Pull the full jug out of the way and place an empty one under the stream. A funnel can be helpful. A makeshift funnel can be made by carefully cutting off the bottom third of a 1-gallon jug and turning this upside down so the neck and mouth point downward—presto!—you have a funnel! A clean piece of cotton can be placed in the funnel to filter out dust and small debris. If you aren't going to use the water right away, cap it and store it indoors.

A bit more practical for the average homeowner is to use a clean 30-gallon trash container. If your house has gutters, simply remove the lower portion of the downspout and place the trashcan underneath where it will collect the rain. Although our house has no gutters, there are several inside corners where the water drains off the roof with a heavy flow. We keep our rain collectors under the cascade, and during a downpour a container fills quickly. We have several containers that we use in this way, and we've sometimes filled three 30-gallon containers during a storm. Because a full container is so heavy, we slide it a few feet to the side as we slide an empty one into place. Always cover the filled containers as soon as possible to avoid an exposed surface where mosquitoes may breed and to protect the water from contamination.

Christopher has also used sturdy 5-gallon buckets with handles and lids to collect rainwater. At one of his former residences, there was a rather large awning. Due to the construction of the house, much of the rain flowing down the roof ran off onto the awning, and a heavy stream flowed steadily from its entire edge. A line of about seventeen 5-gallon buckets at the awning's drip line would fill within thirty minutes to an hour.

In planning a rain collection system, you need to observe the water's flow patterns off your roof in order to position your containers in the optimal locations. Far more rain accumulates during a storm than most of us realize.

The performance of rainwater collection systems over a forty-year period at thirteen locations in California was documented in the publication *Feasibility of Rain Water Collection Systems in California* by David Jenkins and Frank Pearson, published by the California Water Resources Center at the University of California, Davis. The studies cited here confirmed that, quite consistently, 88 percent of California's annual precipitation falls between early November and late March. The rainfall during this wet season is fairly evenly distributed, whereas the scant summer rainfall is highly variable with no observable pattern of distribution.

The authors of this study concluded that rainwater collection would certainly be feasible for the majority of California homes, although this source of water might not be cost-competitive with piped water, due to the expense of storage tanks. Furthermore, acid rain is negligible here; we ourselves test the pH of our rainwater and have never found it to be a problem.

Relying on the rain as your sole or primary source of water is possible only if you've carefully calculated your water needs to avoid any waste and if you have adequate storage to tide you over during dryer spells. Of course, the weather must be cooperative for you to achieve rainwater self-sufficiency.

We've never attempted to rely entirely on rainwater. Our concern has been to save and use at least *some* of that freely falling water from heaven. But the quantities of water available in this way are impressive. We've seldom collected less than 30 gallons in a storm, and on occasion we've collected as much as 400 gallons of rainwater in a single downpour. That's water that we can use as a substitute for the piped water that comes to us from afar.

As for the healthfulness of rainwater, on the beneficial side, rainwater is mostly "distilled water" and therefore contains very few dissolved minerals, unlike water drawn up from the earth. On the other hand, in tests in urban areas primarily in the northeastern United States, rainwater contained lead concentrations equal to or greater than the government's recommended limit for potable water. Microbiological contamination of rainwater was also found, mainly from bird droppings on the urban roofs. For these reasons, Jenkins and Pearson recommend that rain collected in urban areas not be consumed but be used for washing or gardening instead. This is also why we recommend not collecting rainwater until after a half hour of heavy rain so that by then the bird droppings are washed off.

We use our rainwater mainly for the garden and fruit trees and for watering the chickens and other animals. We've used rainwater for washing hair, and it seems to be an excellent hair conditioner, judging by how good our hair feels after a rainwater bath.

If our other supplies have been interrupted and we plan to use rainwater for drinking, we first clean our containers well with biodegradable dishwashing detergent, then rinse them out well before putting them outside. We then cover the opening of the collection container with a fine-weave cotton sheet to filter out particles that wash off the roof. If we forget to add the cotton filter, we'll wait a few hours after the rain stops to let the particulates settle, then siphon the rain out of the container into clean jars for longer-term storage (discussed below).

Due to widespread concerns about acid rain and other forms of airborne pollution, we take the extra precaution of testing any water that we intend to drink. You can test the pH level of rainwater with a strip of litmus paper, which can be purchased at any chemical supply store and at some hobby shops. Where we live in Los Angeles, the rainwater we've collected has almost always been neutral, never highly acidic. Keep in mind that we don't start collecting the rain until after a half hour of a good downpour.

## Water Storage

If you are storing rainwater only for irrigation, you need not be very particular about what sort of containers you use. We have used plastic trashcans, buckets, cans, anything that's available and inexpensive or free.

Of course, it's important to get all that stored water into *covered* containers as soon as possible. If you neglect this step, you'll be breeding mosquitoes for your entire neighborhood.

If you plan to use the rainwater for drinking, we suggest that you start with very clean containers. As noted above, we usually wait a few hours after these containers have filled, let whatever sediment is in there settle, then siphon out the water that we intend to drink. You can make a simple pump to siphon water with nothing more than a piece of tubing, making sure that you don't let the inlet end of the tube get too deep in the water so as to take up sediment from the bottom of the container; start the flow by sucking on the other end of the tube, then put that end into your receptacle. Sometimes, just to play it safe, we run the water through a backpacker's water filter. We have not boiled rainwater, though you might consider doing so if you think that the water may be contaminated.

In general, as a protection against drought or disrupted supplies, you should consider having some kind of water storage system for your household, whether for storing rainwater or tap water. If you live in an urban area, you are wholly dependent on someone else, somewhere else, to make sure your tap keeps flowing. Most of us have no idea where our local water comes from. That makes us vulnerable. You owe it to yourself to find out exactly how your local water gets from its source to your home. You should also investigate and make a list of possible reasons why you might be without water and what forms of contamination might be concerns in your area, and identify all alternative sources of water should the municipal supplies be unavailable. Only by planning ahead will you be prepared for the moment when you turn the tap and no water comes out.

In some communities, a relatively minor electrical blackout (or one of the "rolling brownouts" utilities use for load management) could halt the water flow for some time. Of course, weather emergencies such as drought, hurricanes, tornadoes, and wildfires could disrupt delivery of basic services including water, and human-related disasters such as toxic chemical spills or seepage—or terrorist attacks—can affect the local water supply. In our part of Los Angeles County, any of the many earthquake faults could move and break water mains, and entire sections of the city would suddenly be without water. In most areas of the southwestern and western United States, large numbers of people live in a virtual desert, where the majority of the water comes from afar. It makes common sense to have a few hundred gallons of water stored at your home site "just in case."

Water storage is relatively simple. Water does not "get old" or "go bad," as has often been asserted. Water may begin to taste "flat" with age, but it can be poured back and forth between two containers to aerate it. And though stored water may develop green algae in a few years, this is not harmful and can be filtered out or boiled away so the water can be used for drinking.

Water can be stored in a clean container with a good lid. We have used 1-quart glass bottles, glass jars, plastic buckets, and plastic barrels. Plastic and glass seem to be our main options. Metal containers are generally not used for water storage because the cheap ones will rust, and the good ones (stainless steel) are too costly to be monopolized for water storage instead of being used for cooking and other more temporary tasks. Farmers use large galvanized-metal storage containers for irrigation or animal watering, and these are available at any outlet that sells farm equipment.

But the average person is going to choose glass or plastic (or a combination of both) for backup water containers, because of their ready availability and ease of storage. In our case, we always have some water stored in glass; glass is good because it is inert and will not leach any of its constituents into the water, which may not be the case with some varieties of plastic. But glass is more likely to break. The advantage of plastic containers is that they are very inexpensive and will not break if jostled while being handled or even during a catastrophe like an earthquake. Yet because of concerns about leaching, if you intend to drink water stored in plastic you should take care to use only "food-grade" plastic containers, and even then you should change the water once a year or so.

Is adding bleach necessary? We have sometimes added 5 drops of household bleach per gallon of water before we stored it; actually, when we haven't done so, that was simply out of laziness. Bleach actually has minimal effectiveness for water purification, but will inhibit the growth of green algae in stored water. You should find out what is added to your local tap water by the governing municipality, because the amount of chlorine already added may be sufficient to retard algae.

## WATER CONSERVATION

Water conservation is really the most important part of using water wisely. You can have all the fancy devices in the world for collecting, purifying, and storing water, but in the final analysis, it is the choices that you make that determine whether or not you are a water waster or conserver.

Conservation requires constant and mindful awareness of all the water coming into your urban homestead and flowing out after use. This means turning the water off when you are not using it and making every possible attempt to recycle and reuse. Conservation also means mulching your yard to retain moisture and planting appropriate flora for your ecosystem, because plants that have evolved for very different environmental conditions may require much more water than your climate offers in precipitation. Green-grass lawns in the desert habitats of Phoenix and Reno will hopefully give way to creative and beautiful Xeriscaping with much more intelligent plantings.

Let's consider a number of ways that you can begin conserving and recycling precious water, wherever you live, with minimal investment beyond the effort to change habits.

### Washing Machines

Washing machines require quite a lot of water to wash and rinse a load of clothes. The water level on some machines can be adjusted to the amount of clothes in the load, but this is not so with all models. Many people also tend to routinely wash clothes that are not actually dirty.

One way to reduce the burden on your washing machine is to take one or two small items into the bathtub with you each time you bathe, and then hand wash the items. You'd be surprised what a difference this can make.

We have filled 55-gallon drums with the water from one cycle of a washing machine. After observing how much relatively clean water goes down the drain with every wash, we attempted to find some alternatives. Here are a few possibilities.

Washing-machine drain lines are some of the easiest to convert to a graywater system—an arrangement where nonsewage water from clothes washing, bathtubs and showers, or dishwashers is diverted via special piping to landscape or garden irrigation. Washing machines on porches, in garages, or outside on patios are easy to drain through graywater lines, because you won't need to go through walls. The closeness of the washing machine to your garden or yard is another factor to consider. Because graywater lines usually flow by gravity, it's obviously necessary to have your graywater receptacle lower than the appliance being drained.

At one house where we observed water being recycled, the washing machine was located in a service porch. The residents had attached a 1-inch flexible hose to the drain line and then pushed the hose through a small hole cut into the wooden wall. Five-gallon buckets were kept outside. When the washing machine drained, someone would go out and move the hose from bucket to bucket to fill them, then later carry those buckets to various trees and bushes on the property. Any overflow drained into the ground outside the house. This wasn't an ideal system, because the buckets nearly always overflowed onto a nearby walkway, but the arrangement required hardly any outlay of cash, saved water and energy, and was adequate for that household.

We've also seen a much longer flexible hose connected to the drain line of a washing-machine, long enough to directly reach gardens, trees, and plantings. This seemed to be fairly trouble-free; however, in this case, there were three concerns. One, to reach the yard the hose had to pass over a walkway; therefore the end of the hose closest to the washer had to be kept rolled up outside. When you wanted to use the washer, it was necessary to set up the hose each time by reattaching it to the washer's drain line. Once this step became routine, it took no more than thirty seconds to do, but someone might think the extra step was "inconvenient." Second, the open end of the hose had to be carefully placed, because the first water drained often came out hot. Third, because the house was on a hill, the residents had to be careful not to place the hose outlet at a lower elevation than the washing machine. When they inadvertently did so, all the new water coming into the washer was continually and rapidly siphoned out until someone noticed that this was happening.

Water flows—pressures and drainage—can be perplexing. It's important to thoroughly check out any system you develop, and it may be worthwhile to hire a plumber for assistance if you don't feel competent on your own. Remember that graywater systems are not technically "legal" in most regions, although water-conscious California officials are beginning to approve prototype systems. Eventually separation of normally benign graywater from toilet wastewater may be sanctified by codes, because these systems can conserve and reuse significant amounts of water.

We once had a system whereby the washing-machine water first went into two 55-gallon drums containing water hyacinths, which are excellent natural water purifiers and which don't mind heated water. Once the water cooled, we could open a valve close to the bottom of the drums, to which drain hoses were attached. We'd then move the hoses to whatever plants or trees needed watering.

You need to develop a system to suit your own situation, including the slope of the land, proximity of the washer to the garden or landscape plantings you want to irrigate, and also such factors as ease of operation. A complicated arrangement may amaze your high-tech buddies but might end up being ignored by family members who need to use the system day in and out.

When you are reusing household wash water, you also need to be very careful about what detergents you use, a subject we will consider below.

*Dishwater*

The easiest way to recycle dishwater is to simply carry the dish basin outside when the water has cooled and empty it onto your plants. We have practiced this simple water recycling method for the past thirty years.

On occasion—it might be nighttime, or raining—we empty the used dishwater into 1-gallon plastic containers. We can then more easily carry that recycled water around the yard sometime later. The containers we use for this are old plastic laundry detergent bottles. These are durable and usually have tight-fitting, screw-on lids. Half-gallon plastic milk containers will work but they're more flimsy, become brittle over time, and generally don't have tight caps once they've been opened.

Like clothes washing machines, electric dishwashers can be drained through special pipes or hoses that keep this cleaner water separate from sewage. We have helped to change the plumbing in several houses so that all water going down the kitchen sink would flow into the yard.

Every situation is a bit different. In one case, the drain line was connected to a hose that the residents moved around to different bushes and trees as needed. In another, the water drained into a child's shallow plastic swimming pool that had pinhole leaks, so the water gradually cooled off and then flowed into the surrounding yard over the course of about an hour.

Hopefully, more and more people throughout the country, whether they live in cities, suburbs, or rural areas, will seriously consider modifying the conventional wastewater system to divert graywater for secondary uses.

Did you know that the first navel orange tree was a seedling grown in Southern California by a woman who watered it every day with her dishwater? You can still see that tree today in Riverside, California!

## Soaps

When you begin to take charge of your household environment, suddenly you are forced to deal with the consequences of all the things you'd formerly flushed away, out of sight, out of mind.

When you are reusing graywater to irrigate plants and gardens, you must begin to distinguish between "good" soaps and "bad" soaps. Some soaps, shampoos, and commercial laundry and dishwashing detergents contain bleaches and dyes that will not be good for your plants. Other soaps are completely safe and can even be beneficial to the soil. For example, we have used Shaklee's Basic H for years, and it is excellent in every respect but one: its extraordinarily high price.

Today, with a much broader public awareness that our everyday actions can hurt or help the environment, it is possible to select from a broad array of "environmentally safe" soaps. We're not going to recommend particular brands, because in the past we've recommended some that we liked, and for whatever reason they were unavailable. We suggest that you read the labels on the various soaps and detergents available, and ask questions. We have written letters to the manufacturers when we were uncertain about certain ingredients, and these days it is possible to do consumer product research on the Web. Many companies will respond to customer service inquiries by e-mail or on toll-free lines.

Also, though the food co-ops and health food stores may be a more obvious source of ecological soaps and detergents, don't entirely rule out the supermarket. In fact, you may find some perfectly good ones at the supermarket at prices well below what you pay at special stores.

Be aware that many (most?) of us overuse soaps and detergents. We've heard that the average load of laundry has enough residual detergent in the clothes fibers to permit forgoing detergent entirely for several subsequent loads. While there's no way to prove or measure this, it does seem likely that manufacturers encourage overuse, just as shampoo manufacturers encourage us to "repeat" washings after rinsing, a well-known ruse to encourage consumption.

### Showers and Bathtubs

A widely believed myth that needs exploding is that a shower is a more efficient water user than a bath. Of course, the argument can be made that a quick, two-minute "G.I. shower" will use less water (and thus less energy to heat the water) than a bath. But city folks are city folks, and not one in a thousand will undergo such a rigor without coercion. Especially not on a cold day! You need an iron will to step out of that hot water before you feel good and ready, thoroughly warmed up.

So what evidence do we offer that a bath is actually more efficient than a shower?

1. Bathwater can be bottled and used to flush the toilet, water the plants, and so forth.
2. Bathers can wash three or four articles of personal clothing during each bath, thereby reducing weekly wash loads. This is another reason that we recommend using environmentally safe detergents in your bathwater, because they'll have direct contact with your skin.
3. A small amount of bath oil (or olive oil, and/or fresh lemon peel)

and bath salts turns a plain old tub into a spa, with health-enhancing qualities that a shower can't begin to match.

Even with a steaming hot shower, by the time one steps out, the body has barely begun to sweat. Instead of continuing this healthful, toxin-eliminating process, the body is quickly wiped dry, and talcum powder and after-shower gels and antiperspirants are rubbed on, which instantly plug the perspiration ducts. This effectively seals the poisons into our bodies, all in the name of "looking good." Then we scurry to join everyone else on those nerve-shattering freeways for the purpose of rushing to those places where we will focus eight to twelve hours on "making money."

But this daily cleansing should not be rushed. One hour is the minimum time that should be spent in the tub, considering all the benefits that can be achieved there. Here is the recommended bathing routine we've been taught, known as the Science of Ablutions.

1. Turn on the hot water, catching that first cold water that comes out of the tap in plastic containers.
2. Let the hottest water flow into the tub at a slow, steady trickle and then enter the tub.
3. Listen to the sound of water gently dripping. This in itself is therapeutic to the nervous system; in addition, the water sounds block out the noises of the city. In many ways, the experience of bathing this way duplicates sitting beside a river and offers you an opportunity to do some sustained and deep thinking. Also, by trickling fresh water into the bath, you'll never drain all the hot water from your heater.
4. The bathing room should have ample ventilation, preferably cross-ventilation. Open the bathroom door and/or window about two inches.
5. Add your biodegradable detergent to the water.
6. Lay the clothing to be washed on the bottom of the tub while the hot water is trickling in. When three to four inches of water have dripped in, squeeze each article of clothing two hundred to three hundred *hard* squeezes, underwater. After the bath, these cloth-

ing items should be hung outside where the sun and fresh air will dry them (even on a windowsill, if need be) or if that's not possible on a folding rack in the bathtub itself.

Every pore is an excretory organ. A quick shower cannot properly cleanse or keep open these tiny openings that are so vital to our health, but an hour in hot water can do the job magnificently. A stiff bristle brush is ideal for scrubbing the entire body once the perspiration has begun to flow. This simple scrubbing provides the true "deep cleaning" that many commercial beauty preparations claim but cannot deliver. The hardest scrubbing should be focused on the bottom of the feet, the hands, buttocks, and neck. The scalp must be thoroughly scrubbed as well.

Incidentally, a stiff brush can give an excellent "air bath" if no water is available (on a camping trip, or in the event of a disaster). If you are unable to bathe, use the scrubbing technique to keep your pores open and "discharging." A vigorous dry brushing over the entire body will leave you feeling (almost) as if you'd just come out of a bath.

*DOLORES LYNN NYERGES*

What about recycling tub water? One way to use bathwater "twice" is to give your pets a bath immediately after your bath. We often bring one or even two of our dogs into the tub and give them a good scrubbing. Like most dogs, they initially balk at going into the tub, but they seem to know that they'll feel better when it's all over.

*Fresh from his bath, Cassius Clay has breakfast with his pet kitty Popoki.*

Besides using tub water stored in plastic containers for flushing the toilet, there's another way to recycle bathwater. With the aid of a few good books plus the cooperation of a knowledgeable hardware clerk, you can do your own plumbing retrofit. Go into the basement or under the house, then find and disconnect the pipe that goes from your tub to the municipal sewer system or septic tank. This will usually be 2-inch-diameter pipe. Next, screw an appropriate length of the same diameter pipe onto your bathtub drain line so that the water is redirected out to your garden or fruit trees. If you first draw a schematic diagram, and measure everything accurately before you start, this can be an amazingly simple job. Those who live on a second story can run their bypass pipe over to the second-story outside wall, then attach a hose that drains down to the yard.

The biggest expense—the pipe—can often be purchased used. If you are using galvanized metal pipe, any necessary threading (if not already threaded) can be done at the local hardware store, or you can rent a threading tool. If you do the job with PVC (plastic) pipe instead, you'll need to seal the connections with special solvent glue. The initial cost of PVC is less, yet the life span may be shorter than that of galvanized pipe, especially if any part is exposed to sunlight.

Be sure to install the new drainpipe so that there is a continuous, slightly downward slope to the open end in your garden. Flat sections will slow the water flow, causing a gradual buildup of solids that will eventually clog the pipe. Also, be sure that you bathe exclusively with biodegradable detergents. Many manufactured soaps and chemical cleansers will damage the health of the soil and may kill your plants. As noted above, there are now many genuinely biodegradable detergents.

### The Toilet

Did you know that your toilet will flush even if it is not attached to the regular city water system? This may be surprising, but it's true. If you have never taken a look inside the water tank behind or above the toilet bowl, lift off the top and have a look. The tank is just a reservoir that is kept filled by incoming fresh water. Its function is to suddenly release a large volume of water into the bowl when you push the flush lever, thereby forcing the bowl's contents away into the drainpipe. Fresh water will then automatically refill the reservoir (that's what causes the hissing noise after each flush).

What would happen if you turned off the incoming freshwater supply to the reservoir? The tank wouldn't refill after each flush. Then, the next time you pressed the flush handle, nothing would happen. However, even if there is no incoming water (or if you choose to turn off the supply, which you can easily do with the handle or knob on the intake line, which is accessible behind the base of the toilet), you can continue to flush as long as the drainpipes are intact.

Here's how. First take two 1-gallon plastic containers with handles (such as bleach containers), and using a sharp knife or pair of strong scissors, cut off the entire top so you have an opening that is a few inches across. Try not to cut off or damage the handle. These two containers can be filled with bathwater, then stashed out of sight behind the toilet. When you need to flush, pour the water into the toilet bowl from both containers simultaneously. Believe it or not, 2 gallons is all you need to accomplish what many automatic toilets require 5 (or even 7!) gallons to do. Even if you have one of the new toilets that requires only 1–2 gallons of water per flush, you'd save fresh water by recycling bathwater in this way. You can refill the two containers with water from the containers you fill with your incoming bathwater (the cooler water, before it has come up to temperature, as described above). Finally, to bring the toilet bowl water level back to normal, pour in another half-gallon of recycled bathwater. If you can't save enough bathwater for all your toilet flushes, at least you can make a dramatic reduction in the water used. The extra effort involved will also make your whole family aware of how often the toilet gets flushed, probably some of the time when it's not absolutely necessary to flush.

No doubt there will arise some hilarious moments when visitors first try to use this unusual system. But the rewards come every month on the water bill and in the peace of mind that comes from knowing you're doing what you can to conserve Earth's precious supplies of fresh water— that you're part of the solution, not part of the problem.

If you don't have the space to store all these containers for flushing, or if you can't or don't want to convert your plumbing to a graywater system, you can always carry containers of recycled water outside after each bath and water your garden and fruit trees. If you have a limited number of containers, perhaps a cooperative member of your household could do some garden watering for you while you're in the tub.

## Toilet Paper

Next, let's explore an alternative to flushing toilet paper into the sewers. The founder of the organization WTI purchased a seventy-five-year-old house in Los Angeles and soon discovered that the toilet drainpipe was partially clogged. Because it was prohibitively costly for him to have the pipe cleared professionally or replaced entirely, he innovated several ways of make-do.

Approximately once a week, he'd pour about a cup of glass shards or coarse sand into the toilet bowl and flush it down. The purpose of this was twofold: first, so that the sand and glass would wear away any roots that had grown into the pipe, and second, to scrape away grime and buildup on the inside of the sewer pipe walls. This technique enabled him to postpone the sewer line repairs for more than twenty-five years.

Even practicing this method, however, the toilet would occasionally clog. He saw that this consistently occurred after someone used a large wad of toilet paper. Thus, he came up with an ingenious solution.

Next to the toilet, he placed a sturdy 15-gallon plastic flip-top clothes hamper. In the bottom of the hamper, he put three to four inches of light soil to which was added twenty or thirty earthworms. A sign was then posted next to the toilet paper dispenser asking all users to please throw their used toilet paper into the hamper. People who don't read or don't heed the sign are invariably detected because the toilet almost always clogs and backs up. As the hamper slowly fills with toilet paper, the worms begin to devour it from the bottom. At weekly intervals someone lightly compacts the accumulating paper and adds another layer of soil and worms (along with avocado peelings, which the worms love!).

When filled, the hamper's contents are emptied into a worm farm just outside the kitchen door. This "farm" is a 3 x 4-foot rectangle with a wooden frame at the base. In addition to toilet paper, used tea leaves and coffee grounds, kitchen scraps, vacuum cleaner dust, shakings from rugs and blankets, and sweepings from the daily housecleaning are added to the worm farm.

This hamper has "earthy"—but never offensive—odors emanating from it, somewhat like the potting area of a nursery or small greenhouse. These odors can be detected momentarily when the lid is swung aside to deposit materials, but they don't linger in the bathroom air. In fact, there's nothing to indicate to a visitor what truly marvelous activity is taking place except a small card mounted on the lid with a listing of the dates when various maintenance tasks (for example, adding more worms) have taken place.

## Toilet Alternatives

We had to chuckle when we heard of a family who decided to shorten the length of their "survival test" because their portapotty broke. These folks had wanted to see how well they'd survive in their urban home if an earthquake severed all utility lines and truck routes to the city. So they turned off their water, their gas, and their electricity and used only the food, water, candles, and other supplies they had stored in the home.

Yet they called off the experiment when their emergency toilet ceased working properly. Why did they stop the test when the hardships were just becoming real? After all, when a crisis is suddenly upon us, we'll have no time for buying amenities—we'll simply need to use what is at hand.

The subject of human waste disposal is probably the most sensitive and most ignored area of disaster planning. In our sophisticated, modern society, we evidently don't want to bother with such matters. We use the toilet, flush, and forget. Let someone else worry about it.

And yet, properly composted human excrement can be a valuable source of fertilizer and even burnable gas, as the Chinese have proved. As of the early 1980s, there were 7.2 million methane digesters in rural China that used human and animal wastes to provide fuel and fertilizer for about 35 million rural people.

Let's consider some alternatives to conventional indoor toilets.

RV TOILET. Once for a period of three months in his Los Angeles home, Christopher exclusively used one of those simple, "bucket in a bucket"-style toilets designed for a recreational vehicle—a "dry" toilet, not a chemical-fluid unit. The test's purpose was to ascertain the practicality of relying upon such a toilet after an earthquake. When it was full, Christopher emptied the bucket into a trench in the yard, which was then layered with straw, worms, and worm castings. Eventually, he planted tomatoes in the trench, and ate them a few months later. (According to numerous sources, tomato plants are capable of growing in—and processing—sewage.)

The RV toilet was not difficult to use. The real challenge involved finding an effective and ecological method to combat the unpleasant aroma. After various experiments, he found that the most practical and economical method for eliminating the toilet odor was pouring some lemon juice and baking soda into the toilet after each use.

WORM-COMPOSTING TOILET. Three years later, Christopher conducted another test, this time in the back yard. In a secluded spot, he set up an outdoor toilet, which consisted of the toilet seat from a hospital-style potty and, instead of the usual pot under the seat, a large wooden box. This box was sized to fit under the toilet seat. Into the box he placed a layer of worms and worm castings and some partially decomposed straw. After each use of this toilet, the excrement was covered with another layer of worms and worm castings. This system proved to be simple, odor-free, and fly-free. The key to its success was the addition of earthworms. Christopher used the type of earthworm known as redworms, which (as explained in chapter 5) are the most common variety bred in backyard worm farms because they are rapid breeders and can tolerate broad fluctuations in temperature. Earthworms are great partners in the composting process: they continually burrow and digest organic matter, breaking it down into nitrogen-rich plant food.

After the box under the toilet became full, it was moved to the side to give the earthworms time to process the contents, and the toilet seat was placed on top of another wooden box. Two months was adequate for full decomposition of all excrement and toilet paper in a box. The contents of each seasoned box—a rich, loamy soil full of earthworms—were distributed around the base of fruit trees.

The idea for this worm toilet came to Christopher while he was observing the earthworm farm he had established under one of the rabbit coops. As described in chapter 5, the rabbits were in suspended cages that had a screen-mesh bottom, allowing their droppings to fall through. Using old lumber, an enclosure was constructed on the ground under the rabbit coop. Earthworms proliferated rapidly in this mix of rabbit droppings and urine, and there was a conspicuous absence of flies and odors around the rabbits. This was an ideal arrangement, a virtually maintenance-free worm-composting operation that greatly reduced the necessary cleanup of the rabbit hutch.

We had another motivation to pursue development of the worm-composting toilet, for reasons besides earthquake preparedness. Some estimates indicate that about 50 percent of the average urban house-hold's water is used to flush the toilet. Anyone who is concerned about this huge waste of water, as well as the waste of a potentially valuable fertilizer, has probably investigated commercially available composting toilets.

There are several types of manufactured composting toilets now available through catalogs and even in home supply stores. Some of these use electricity for fans and heaters, and others do not. The electrical features are usually recommended for smaller models, because these units require more active management, and the fans and thermostatically controlled heaters help to aerate and dehydrate the contents of the holding tank. Larger models, on the other hand, generate enough of their own heat to fuel the decomposition process.

Certain of these models can be installed in an urban bathroom and will look like a slightly elevated toilet. Other models are larger, designed for rural or wilderness settings where there are no sewer lines or where septic tanks are uneconomical, and where space is less of a challenge.

We have looked at several of these composting toilets and can see that some of them are quite promising. But even the least expensive of these costs around $900. Thus, we attempted to find an easier and cheaper method that would nevertheless conform to standards of health, cleanliness, and ease of operation.

Our worm toilet is an ongoing project, and as we continue to experiment, we can see already that the potential for such systems is enormous. Not only could these toilets save vast amounts of water, if set up properly they could produce economically valuable compost, and they would be entirely viable even after a disaster such as an earthquake. In wilderness areas, such as National Parks and at remote cabins, this kind of toilet makes far more sense than the currently common outhouses, the contents of which are considered "waste" at best and a health hazard at worst.

The toilet alternatives we've shared here will save water, but keep in mind that these ideas are experimental and controversial. They are important to know, especially for emergencies, but if you plan to implement any of these systems, check your local regulations and proceed with caution and common sense.

## THE KEY TO SUCCESS: CONSERVATION

The whole subject of water use and conservation underscores the fact that our choices in so many countless little areas may result in broad, cumulative ramifications in the world around us, many of which we never see. We believe that the only meaningful "revolution" is a personal revolution, whereby we choose to live our lives reasonably, rightly, and lightly on the earth.

Ecological living can and should be practiced in the cities. To do so, we must value self-sufficiency as much as—really, more than—convenience and be willing to make a series of minor adjustments in our accustomed wasteful way of doing things. A resolve to conserve water is great way to begin creating a better way of living.

# 7  Homestead Energy

## Appliances, Heat, and Electricity

*I doubt that any of us knows where he is in relation to the stars and the solstices. . . . Like the wilderness itself, our sphere of instinct has diminished in proportion as we have failed to imagine truly what it is.*

—N. Scott Momaday, from "An American Land Ethic"

We all interact with different forms of energy and power constantly, most of us ignorantly. Nevertheless these interactions determine the quality of our lives.

Here in the first years of a new millennium, with unpredictable oil prices and supplies, all kinds of "experts" continue to glumly tell us that there are no other viable sources of energy than fossil fuels. Those who believe these "experts" continue with business as usual. Meanwhile those who are aware that practical, affordable solar-based alternatives have existed for decades are already pioneering a new way of life.

In our home, we have long eschewed many of the "modern" appliances that now are ubiquitous. It seems now as if there is an electrical gadget for every yard and kitchen task that in the past would have been done by hand or with hand tools. Fortunately, many of those durable old manual tools are still with us, and in many cases they do the job just as well as or better than their energy-hog counterparts. So whether you use hand tools by preference or as backups in a pinch, there's no real need to be helpless simply because the power goes out.

For example, we got interested in old-fashioned, hand-crank coffee grinders more than a decade ago, and have purchased many at flea markets, for our own use and to give away as gifts. We have found one particular style to be the best—there is a large box with a hopper inside, and you can set it in your lap while you grind. Truth be told, we also have a little electric coffee grinder that grinds enough grounds for a pot in about fifteen seconds, but we still prefer to grind by hand. We'd have no problem grinding our beans if we lost power.

We have several manual can openers, because we don't see the point in having an electric device to do such an easy job. We also have a few different manual citrus juicers. We have used the electric type, but they don't seem to extract as much juice as the good manual ones. We strongly recommend against electric knife sharpeners because they cut away fine slivers of metal, which gradually ruins high-quality knives. We always sharpen our knives by hand.

One can take great personal satisfaction in using hand tools where possible, realizing that so many of the "essential" modern appliances are by no means necessary. In many cases, in an emergency you can simply do without whatever requires power. We have been in homes with countertop devices for many different kinds of food processing, and we often don't even recognize the function of these gadgets.

You need to decide for yourself what household uses of energy are really important.

For instance, we find the television to be a most remarkable tool for showing us what is happening far away. We remember when we had no power here after the Northridge earthquake occurred on January 17, 1994. We were not concerned about lighting (we had candles and flash-lights and oil lamps) and we had no problem with cooking (we have a fire pit in the back yard) and we weren't even especially concerned about the food in the refrigerator. However, we really wanted to know what was happening in other parts of town and beyond, but we had no radio or television. In time, we purchased several radios that do not re-quire batteries—those remarkable BayGen crank radios, which provide about thirty minutes of radio play for each minute of cranking. We also bought a battery-operated color television so we could watch the news reports. This requires ten D-cell batteries; a black-and-white model, if you could find one, would use less power. We regard our crank radio and battery-operated TV as wonderful inventions, and we're glad we have them.

On the other hand, because this was a duplex, there were two electric dishwashers in the house when we arrived here, and we removed both of them. We find dishwashers to be a waste of electricity and water, all in the name of "convenience." Of course some people will debate this point, but we believe that you can get dishes cleaner by hand, with less water and less energy for water heating.

And as you already know from the enthusiasm we expressed for kitchen scraps in our discussion of composting (see chapter 5), we consider garbage disposals to be a technology invented to keep plumbers employed. Along with breaking down frequently, these dreadful machines waste water and electricity, and all the wonderful organic "refuse" that they chop up and flush away is far better fed to the animals or put into a backyard compost pit or worm farm.

"Living lightly on the earth" is a state of mind, an inner commitment to take action personally to be part of the solution to the enormous problems facing us today. We feel that such solutions must always begin in the home. We have met too many outspoken people with great ideas who do not practice what they preach. We are attempting to put our ideas to good use, one by one, in our own private home and personal lives.

We still have a long way to go. We want to install more windows and some skylights that will allow us to work indoors during the day without electric lights. In a climate that requires wintertime heating or summertime cooling, investments in improved insulation and weatherization (caulk, weather stripping, door and window seals) can make a huge difference in energy savings. And we are beginning to install a simple, introductory photovoltaic system, as described further on in this chapter, using electricity-producing solar modules to supply an increasing portion of our household energy with free sunlight.

Let's summarize a few of the key steps to household energy efficiency and even independence:

1. *You can do without some electrical devices.* This will probably involve changing your behavior, for instance, thinking twice before switching on an electrical tool or appliance when a nonelectric alternative will work just as well or better.
2. *You can learn to use your existing devices more efficiently.* This step, too, requires changes in habit, but once you've understood

the extra expenses caused by inefficiency and waste, you'll feel good about it—plus you'll save money by practicing efficiency.

3. *You can purchase new appliances that render your household inherently more energy efficient.* This step requires initial outlays of money, and in some cases higher short-term expenses, but with certain especially wasteful appliances the best way to save energy and money is to immediately replace the old, wasteful model.

Next let's look at specific strategies associated with each of these approaches.

## 1. *You can do without some electrical devices.*

It is entirely possible to live well without any of the following, none of which seem to us to be really necessary:

*Weed whackers.* These can kill trees if you prune too close around the base. We've used them on occasion to cut down foxtail grasses and other pernicious plants, but rarely do we whack down our "weeds," many of which are actually useful and edible plants.

*Lawn mowers.* Well, as you know from chapter 2, we don't have a lawn, but even if we did, we would cut it with a hand-pushed rotary mower. There are good, reliable, self-sharpening models available from most garden centers and many catalog companies.

*Leaf blowers.* These are noisy, fume- and dust-spewing contraptions that some so-called gardeners claim are indispensable. A rake and a broom work fine.

*Hair dryers.* What's wrong with a towel and fresh air?

*Electric toothbrushes.* Really?!?

*Electric can openers.* Ever hear of people who couldn't open a can after an earthquake or storm because their power was out?

*Electric knife sharpeners.* As noted above, these destroy the temper of good knives.

*Electric citrus juicers.* An utterly superfluous use of electricity.

*Microwave ovens.* Are we all in such a hurry to eat? Also, these units are mainly designed to cook highly processed "fast foods," which are expensive and less nutritious than whole foods.

*Automatic dishwasher.* Wash them by hand.

*Garbage disposals.* Make a compost pit!

*2. You can learn to use your existing devices more efficiently.*

*Lighting.* Simple—turn off the lights when you leave a room. This easy act is wholly within the realm of anyone who has a modicum of personal discipline. Also, in this era of "cheap" (but short-lived) incandescent lightbulbs, people consider compact fluorescent bulbs pricey. But slowly the message is getting out that fluorescent bulbs last ten thousand hours or more and provide equivalent light with a fraction of the energy. Ever wonder why incandescent bulbs get too hot to touch? Instead of light, they are wasting a large portion of their electricity as heat. Swapping all of those wasteful incandescent bulbs each time one burns out with a durable, long-lived compact fluorescent bulb is one of the single best ways to save energy at home.

*Television.* Don't let the TV or radio run all day as "background noise." Why, oh why, do so many of us do that? Some say that one cause of this uniquely modern habit is the fear of being alone with oneself, "listening to oneself think," as it were. That constant chatter of the TV in the background can actually keep you estranged from your own inner self.

*Refrigerator.* Plan your trips to the refrigerator so you can open the door, get what you need, and quickly close the door. You can keep the outside covered with sheets of insulation (such as those bubble sheets available from pack-and-ship shops) to increase its insulation value. You can also keep the freezer full, because it works more efficiently that way; instead of needing additional energy to cool empty space, the mass of frozen goods retains coolness. Also, with a brush or vacuum cleaner, clean the dust off the back coils so excess heat can radiate out and away.

*Stove.* With a gas range, use the smallest gas flame possible and keep all pots and pans covered while you cook. This not only saves gas, but reduces scorching of pots and pans and produces food that is better for you.

*Heating and cooling.* These can represent a big portion of one's monthly utility bills. It often seems that we have lost the ability in industrialized cultures to experience the actual conditions of the real world—times of heat, times of cold—even though the body has evolved (and clothing has developed) to deal very effectively with a wide range of temperatures. Learning to deal with greater extremes of heat and cold involves some mental conditioning and some physical conditioning and cleansing. Your pores actually help to regulate your sensation of heat or cold—yes, folks: our skin, if clean, is the body's best temperature regulator.

If reconditioning your physical responses sounds like too much of a challenge, your immediate course of action should be simply to regulate your clothing according to the temperature. Put on a sweater and a second pair of socks before turning up the thermostat. Then, if you must use the heater or cooler, be sure to open or close doors and windows, as appropriate, so you are not simply wasting more electricity (or gas) than is necessary. Be sure to clean dust out from coolers and heaters annually, because a buildup of dust can decrease their efficiency.

*Water.* You may not think of the water supply as requiring energy, but unless your water arrives entirely by gravity feed, pumps and pressurization are involved. A variety of water-conservation measures are discussed in chapter 6, but the importance of saving water can't be overemphasized. We now know that water is a finite resource. As more and more people reside in crowded urban centers, the sources of water are severely overtaxed. Many of the cities of the West and Southwest could not exist if water was not being pumped in from afar. Case in point—Los Angeles! When you increase the people, and decrease the water, you will have problems of gigantic proportions, somewhere, sooner or later. Treat every drop as precious. If you have a faucet leak, fix it. Turn off the spigot when you are done using it.

### 3. *You can purchase new appliances that render your household inherently more energy-efficient.*

Anytime you need to purchase a new appliance, make the effort to get the one that is rated most energy efficient and most recyclable at the end of its usable life. The government's Energy Star program with its distinctive logo makes it easy to identify and evaluate energy-efficient appliances, including furnaces, air conditioners, heat pumps, refrigerators and freezers, dishwashers, clothes washers, light fixtures, as well as audio-visual equipment, computers, and office equipment. The bigger and more energy-intensive the appliance, the more incentive you have to replace a wasteful model with a new, energy-efficient one. This may mean a higher initial purchase price but will almost always mean savings in the long run. It is always a false economy to buy what is cheap and expedient now, then have to pay and pay as the months and years go by. The book *Consumer Guide to Home Energy Savings* (updated each year; see the bibliography) includes listings of the most efficient products available.

*Water heaters.* There are some excellent models available. You can also add extra insulation to an existing water heater. Gas is more economical and efficient than electric.

*Refrigerators.* This is one of the biggest power uses in the modern home, so it pays to shop around for a model that will provide you years of good service at a low monthly energy cost. An old refrigerator is a gigantic energy hog. New federal efficiency standards went into effect in 1993, and some current models are over 20 percent more efficient than required by the standards.

*Stoves.* Gas is more economical and efficient than electric models. As with an old refrigerator, you may be better off replacing an inefficient old range-and-oven combination, and sometimes a local gas company will give you a rebate in order to encourage you to convert. (In addition, we are big fans of solar ovens, which are discussed in more detail below.)

*Solar electricity.* For people in many parts of the country, it may be cost-effective to install photovoltaic panels on your roof to produce some of your own electricity. Solar equipment is a far more affordable and more viable option than most people realize, even for city dwellers. And if you could reduce your electrical use 50 percent by wise use and nonuse, you would need to spend that much less for a solar-electric system. Depending of course on your motivation and self-discipline, this level of potential savings is wholly within reach of most people living in typical households. We suggest you start small so that you get used to some of the monitoring and conservation routines that are involved in harvesting solar electricity; before long, these become second nature. We will describe our own adventure with solar electricity later in this chapter.

Overall, we have chosen to adopt a manner of thinking, and a manner of acting, as if we were living in an self-contained ecosystem and each of our actions potentially affected our next glass of water, our next breath of air, our immediate food supply, and so forth. That already is the case, really; and how we interact with the land, air, and water determine the quality of our life. And yet in a city, many individuals tend to shuck their personal responsibility because the consequences are not so immediate or apparent. They can always say "Look at what the other guy is doing," or "Why should I conserve when they don't?" or "What difference can my puny efforts make when there are so many people here?" Unfortunately it is this type of thinking that exacerbates our problems.

Where possible, we have chosen to adopt low-tech modes of behavior and low-tech solutions. George Sibley, author of *The Desert Empire*, states (in Sibley's Law #1) that technological problems increase in exact proportion to the complexity of technological solutions. In countless cases where bureaucratic high-tech measures were employed, the "solution" to a problem has become a new problem in itself. A current example of this phenomenon is the use of genetic engineering to modify plants and animals to be more "productive" according to the dictates of industrial agriculture.

Overreliance on technological solutions is another fancy way in which we've learned to shift our responsibility away from the individual. The home is the workshop for testing real-world solutions. If you can't use a given practice to make a difference in your personal life at home, it probably isn't a practical solution with wide potential. Moreover, if a solution keeps you distant from the impacts of your actions and disinterested in their consequences, it is not a real solution promising widespread adoption.

Now let us consider specific areas where urban homesteaders can save energy and become more economically self-reliant in the process.

## COOLING

Although we actually do have an electric air conditioner mounted in the window of our front living room, we rarely use it. We found that our home has poor air circulation, partly due to the fact that our structure is L-shaped, with the long leg of the L blocking the dominant breezes, and partly because our home is sited down lower than the surrounding hills.

This has meant that when it is very hot, we have not had the advantage of natural air circulation.

We came up with two remedies that have helped us work *with* rather than *against* nature.

First of all, we painted our roof white and were amazed to find that right away we experienced summer temperatures at least 15 degrees cooler. In fact, when we originally painted the roof white, we didn't realize we'd be getting a cooler home. The roof was worn out and there were many holes, resulting in rain leaks. We simply couldn't afford a new roof and found a "liquid rubber" product that seemed like a cost-effective alternative as it promised to seal the leaks and make the house cooler as well, because it was white. At first we weren't particularly thrilled about a white roof, because this sealant was a really *brilliant* white. We would have preferred a light earth tone, but the product came only in white, so white it was. We figured every helicopter in L.A. County probably spotted us from miles away.

The liquid rubber didn't really stop *all* the leaks until we had about three coats of it up there, and even then not all the moisture was blocked. However, for the four or so years that we had that white roof, the beastly hot times of summer when the outside temperatures were in the high 90s and 100s were slightly more bearable in our noticeably cooler house.

Eventually, we did have a new roof put on, and as a result of our previous experiment, we opted for the whitest shingle that was available.

Our second remedy, in order to keep the house naturally cooler in summer, was to install metal security doors with heavy-gauge screening at both ends of the house. These have allowed us to keep the main doors of the house open, providing cross-ventilation thanks to an airflow throughout the cooler hours of night. We had no idea how much more comfortable the house would be when we were able to allow cool air to naturally flow through!

## HEATING

Our heater is a wood-burning fireplace, not an electric or gas heater.

We both grew up in homes with fireplaces, and we remembered

how the fireplace was the warm center of the home during the winter months. Hearths seem somehow essential to making a house "home." Although our Los Angeles home didn't have a fireplace, we decided that we would eventually buy or build one for our living room.

In the meantime, we lived for years without "central heating." A wall-mounted natural gas heater in the rear of our home never worked right, and the repairman wanted too much to fix it, so we just never used it. Hearing that we had no furnace in the house, friends and family often responded with concern, even shock, which we both found humorous. After all, we live in Southern California, not in Buffalo. Heating is simply not required most of the time. Keep in mind that the Native Americans who resided here more than five hundred years ago wore just about no clothes most of the time. They didn't need to! We've found that we can sit outside in the sunny back yard and do work there, even in winter.

Because of its layout, our home seems to be cooler inside than the air outside the house during the winter; yet we found that we grew accustomed to the unheated air temperatures, and if it gets really cold, we simply put on another sweater.

We did buy one of those small electric space heaters with a fan, but stopped using it because it drove up the electric bills noticeably. We also purchased an energy-efficient "sealed burner" oil heater and still use that on occasion for space heating.

But all along, we really wanted a wood stove or wood fireplace, so we kept looking. We were not particularly interested in a gorgeous antique that would impress the visitors, nor did we think that we should pay thousands of dollars for the privilege of burning wood. We just wanted a well-built workhorse that would cook our food and warm the house.

Over the years, at flea markets and yard sales, we purchased three different wood-burning stoves at bargain prices, but for various reasons, none of these was quite right. Then one day Dolores found a a freestanding fireplace available at a yard sale. We drove about a mile away to a house in an older part of town. Apparently the folks were moving, or cleaning out things they no longer needed. There, out in the driveway, was a freestanding sheet-metal fireplace designed to sit in a corner. It was triangular in shape, with a large firebox capacity and a screen covering the front so you could see the fire burning. It had also been modified to be supplied with a gas line, which we thought was an added bonus in case we wanted to use natural gas instead of (or in addition to) wood.

We looked the fireplace over eagerly yet carefully, to see if it would serve our needs. We'd never seen a fireplace exactly like this one before, but we'd seen enough that we knew generally what to look for. We purchased it for $70.

For the first few months, it sat in our garage, that repository of dreams and great intentions. At last we decided it was time to actually get the fireplace installed. We spent about a month figuring out what was needed. At one fireplace shop, the owner—although he was very polite and helpful—wanted to sell us the wrong piping, which he assured us was exactly right. We nearly bought it, but due to our penchant for always seeking the best buy, we held off getting the pipe that day.

In general, there are three types of chimney pipe arrangements that are safe for freestanding fireplaces and wood stoves, depending on your situation. There are the pipes that go through a wall to the outside. There is the setup that goes directly out the roof—used in cabins where there is no interior ceiling. And then there is the type we needed, where you have both a ceiling and a roof to go through. It's lucky that we checked around and asked lots of questions, because we otherwise would have spent twice as much and ended up with a great headache.

Unfortunately, you never know what questions to ask until you are well into a project. In our case, the sheet-metal fireplace was built by a small, now defunct shop. No one today, we were told, made models with 9-inch openings, and our corner fireplace was somewhat rare (though not particularly valuable, money-wise).

We finally decided to go with 8-inch piping, and returned to the original fireplace shop, now that we knew what we wanted. The price was nearly $700 for all the necessary materials. We needed to have stovepipe to go from the fireplace to the ceiling; a special insulated box that fits into the ceiling rafters; a section of double-walled pipe that would pass through the roof and extend up above the peak; a spark arrester for on top; and a little metal skirt that supports the base of the exterior pipe and keeps it from making contact with the roofing.

Reluctant to pay $700, we went home and called a regular hardware store, the likes of which often sell such equipment for less than fireplace shops. It was immediately clear that these folks got very few requests for fireplace smokestacks, that it would have to be a special order, and that they didn't want to bother. We talked them into giving us the name and number of their supplier, and after two more phone calls, and a drive a little farther from home, we purchased everything we needed for about $350.

We were of course very happy, and "temporarily" stored all the parts in the garage. Next we put the wood fireplace, already in the corner in our living room, onto a new fireproof platform.

Then we contacted a local metal shop and had them custom-make a stovepipe "collar" that would join the 9-inch-hole to the 8-inch stovepipe. This part cost $59, and it took the shop a month to get around to finishing that little part.

Finally, nearly a year after we'd purchased the fireplace, we decided "let's get it done." We had spent more time thinking about the details of the project than it actually took to do the real work.

The first challenge was figuring out where the holes would be cut, which required going up and down a ladder into the attic in order to locate those holes in precisely the right place. Once it was clear where the fireplace would go, we cut a square hole in our ceiling for the box that supports the double-walled section, and another hole directly above it through the roof. The roof hole had to be round, exactly 1 inch larger all around than the smokestack. The cutting took about an hour.

The holes were cut, the box was secured to the ceiling with nails, and from up on top of the roof the smokestack was set into place with its spark-arrester top; then we had to secure the protective metal skirt that goes around the chimney stack and seal the seam with tar to prevent leaks. These steps all went very quickly.

Finally we had to fiddle with the various sections of pipe running from the top of the fireplace to the ceiling. This took a little finesse, and we both had to work on it together, because the piping went straight up and then bent at an angle over to the opening in the ceiling. At last, we pushed the fireplace a little to make sure all the pipes were seated snugly, and we were done!

It took awhile to clean up and to inspect the work again—and again—to make sure we hadn't forgotten anything that might cause the house to catch fire.

That night, in the winter chill of January 8, 1998, we had our first fire. This was a very glad occasion, more than a year since we'd bought the stove.

Our freestanding fireplace has completely transformed our home. We would strongly encourage anyone without one already to seriously consider installing one. As mentioned above, on very cold nights we had been using those small electric heaters that really drive up your electric bill. The fireplace made the house really feel like a home, and we now are uncertain how we got along without it.

We were amazed at how well the fireplace heated most of the house—not only the living room—with just a few logs. We remembered that built-in fireplaces never heat a room very effectively, and in fact, conventional fireplaces suck warm air out of a house and send it up the chimney. We concluded that this freestanding wood fireplace does a better job for several reasons: first, it is set out in the room about two feet; second, because this one is a corner model, the heat radiates off both walls, outward to the entire room; and third, in addition to the firebox itself we benefit from the heat of the six feet of exposed stove-pipe inside the room.

In our case, the transition to wood heating was fairly easy, because we had plenty of firewood readily available. We were actually doing a neighbor a favor by cleaning up and carting off large amounts of dead and fallen wood from his property. He had a great number of very tall eucalyptus trees, with many dead branches in need of removal. Our first season of firewood came entirely from our weekly cleaning of his yard, just for the cost of our labor. How's that for a win-win situation?

We also became aware that city folks don't value firewood, presumably because fireplaces are now a thing of the past. This means that we can drive down the street and find stacks of good firewood set out for the trash collector to take away. We have filled our truck many times with discards from other people. Sometimes we need to cut or split this salvaged wood, but other times we find neat bundles all tied up to be taken away. We have also found it advantageous to make friends with some local tree pruners, who have occasionally provided us with whole truckloads of quality firewood. Our taking this wood saved them the "tipping fees" they would have had to pay to drop it at the city dump! Another win-win situation.

It is easy to see why a wood-burning fireplace or wood stove is synonymous with self-reliance and homesteading. There, all wrapped in one, you have a heater, a cooker, and the "caveman TV." If you can stay warm without having to pay exorbitant fuel costs, you're on your way to self-sufficiency. And if you can cook your food without gas or electricity, well, you are really on the right track. Plus, whenever we have sensitive papers that we don't want to just toss into the city's recycling bins (who knows who may see them?), we simply use them as tinder for the evening's fire.

*A word of warning.* Homes burn down all the time, due to gas leaks and explosions, fireplaces, wood stoves, Christmas trees, and so on. Don't cut any corners when it comes to installing your stove or fireplace safely. You want to stay safely warm *inside* your home; you don't want to be outside, watching the house go up in flames because you tried to save a few dollars somewhere. And you don't want to suffer from asthma and other breathing difficulties because your wood-burning heater is polluting your living space.

Although it is widely believed that wood "burns clean," this is not altogether true. While wood smoke smells good, burning wood can produce up to 175 different combustion gases. And no fumes from combustion are completely benign, whatever is being burned. For instance, pines and other conifers emit hydrocarbon gases when burned, which means technically that these trees produce smog. This is actually true whether or not the trees are burned, because in very hot weather, the gases evaporate out of the needles and cause the haze you see in certain wilderness areas, such as the Blue Ridge Mountains and the Great Smoky Mountains (which is where those names come from). By contrast, according to an article in *Popular Mechanics* (January 1979: 7), some of the hard plastics—specifically certain polyvinyl chloride compounds—produce no chlorine, no phosgene, no nitrogen dioxide, nor sulfur dioxide or ammonia when burned. This surprises many people—to learn that some kinds of wood produce more harmful gases than some plastics.

In any case, you want a properly vented wood stove. When using the fireplace we never close our windows all the way, so we always have an influx of some fresh oxygen.

Also critical, with any stove installation, is to clean out the stovepipe (especially horizontal sections) on a regular basis, because creosote and other particulates will tend to collect there and can ignite suddenly.

Go into a project like this with determination to do an impeccable job, and your wood stove or fireplace will be a rewarding part of your homestead, whether urban or rural. Needless to say, what we're offering here is merely an account of what we did in our situation. There are many ways to safely install wood stoves and wood fireplaces, if you are handy and if you have a good sense of what is *safe* and what is not.

For example, when Ernest and Geraldine Hogeboom were living in Pasadena, California, they kept a potbellied stove right in the middle of their living room. Ernest installed the stove by simply running a pipe up to about a foot from the ceiling, and then horizontally all the way to his window. The horizontal pipe was held up every few feet by a wire that hung from eye-screws secured to the ceiling. Ernest never purchased any "code-approved" materials for sending the pipe through the wall. He put a piece of plywood in the window—plywood wrapped with several layers of aluminum foil—and cut a hole in the plywood to fit the pipe. We're sure the Pasadena fire department would have taken a dim view of his makeshift vent. Also, had there been a house fire, there is the likelihood that the insurance company would have brought up the issue of an improper vent. But they never had a problem or a fire in all the years they lived there.

Ernest and Geraldine are old friends of ours whom we met through WTI. They "homesteaded" at their place in Pasadena, keeping goats, chickens, bees, ducks, and a huge garden. It was really something to visit. But they had "growing pains"—they needed more room to do what they were doing—and moved to Northern California where they found work at a dairy farm.

Back in the early 1970s, when we were just starting our Wild Food Outings and when the Survival Training School was in full swing, Ernest Hogeboom and Richard White used to show trainees a unique fire starter for either a home fireplace or campfire.

Into a paper grocery bag, they would pack as much discarded paper, cardboard scraps, and toilet paper tubes as possible. They would use different-size bags for different-size fires. They would also put into the bag a little bit of the hard plastic, the translucent type of plastic that was used then for some milk containers. Clearly, this helped to get the fire going, as the plastic would burn very hot. Someone would always ask, "Isn't burning plastic bad for the environment?" and they would explain that no material burns entirely cleanly, and that, contrary to popular opinion, there are actually some plastics that burn cleaner than wood. They weren't particularly advocating burning great quantities of plastic, but they did want the students to consider the question in perspective.

Another really effective, really simple way to make kindling is to take several pages of newspaper, roll them up diagonally like a big tube, then tie the two ends in a knot: the ends ignite readily, and the knotted part burns very hot.

## COOKING

If the gas or electricity goes out, many folks have no clue how to cook their meals. Our regular stove for cooking is an old natural-gas kitchen range, and of course we also have the wood-burning fireplace described above, which can be used for cooking if necessary. Admittedly, our fireplace has a small, flat top—wide enough for one pot at a time—not really enough space for any serious cooking. Although we've heated some drinks and other small dishes there, the fireplace's purpose is really heating the house. Even so, in an emergency we would not be without means of cooking.

We would never consider buying a microwave oven; we've never felt comfortable with the idea of cooking our food with high-intensity radio waves, and microwaves seem like a high-tech, potentially unhealthy toy compared with a simple flame, which will do the same job well.

As we live in Los Angeles, not a day goes by that we don't marvel at that great wonder of technology, the common kitchen stove, that allows us to boil water, bake bread, sauté eggs and vegetables, and otherwise prepare hot meals with the turn of a dial. We'd be the envy of nearly anyone living anywhere on the planet a hundred years ago. And yet this marvel is so common that it is largely taken for granted today, and we are so dependent that we've lost the ability to cook without modern utilities. What a shame!

While anyone would be inconvenienced if they had to suddenly shift from gas to some other less modern method of cookery, everyone should be able to revert to the caveman essentials and know what is necessary to cook a meal. In our case, we can always go out to our back yard, build a small fire, and boil water, make soup, or cook breakfast. All that's necessary to do this is a circle of bricks or rocks, a grill, some wood, and a way to start a fire—nothing difficult or complicated.

We have occasionally cooked over an open fire out back when we've wanted to do so. We had another wood stove out back for a while, and cooked on that. We had no smokestack, but didn't need one because we were outside. In addition, we have had a variety of wood-burning devices with which we've experimented. These have included hibachis, large cans, old barbecues, and a unique backpacking stove called a Pyramid, which is small, portable, and extremely efficient. All of these are easy to use and have amply illustrated to us the ease of cooking food without a conventional kitchen stove.

## Solar Ovens

*In every country, one wonderful solution comes up every morning.*

—Anon

We have also invested considerable time and some money experimenting with the many ways to cook food with the sun. Just imagine! You can actually take scrap materials—old cardboard boxes, newspaper, a few short cans, plastic, or glass—and create a device that cooks with free sunlight. Using the sun as much as possible is always less expensive and less harmful to the environment than using gas or electricity for cooking. Plus, although there's no way to prove this, we believe that food cooked with the sun not only tastes better, but *is* better.

We have a solar oven built from plans found in the early 1970s in the *Mother Earth News*. This oven, which has served us well for more than twenty-five years, is a fiberglass-insulated box with a sloped pane of glass on top, set at a 45-degree angle. We have seen these made of wood, but we made ours of sheet metal because that's what *Mother Earth News* recommended, and it was therefore easy to rivet the pieces together. Food is put in and taken out through a little door in the back. We once had rigid aluminum reflectors on this oven, but found them more trouble than they were worth, difficult to keep in a set position, and so we removed them. Even without the reflectors, on sunny days we typically get temperatures of 250 degrees Fahrenheit and above, and this is fine for our style of cooking. Usually we use the solar oven for making breads and biscuits, and sometimes for cooking vegetables.

We also purchased a commercial solar oven, the Burns brand Sun Oven. This has basically the same design as the oven in the magazine article, but the commercial one is constructed better than our garage-made model and not only achieves temperatures of 350 degrees Fahrenheit but also maintains that concentration of heat better. (Note that we do not recommend cooking foods at temperatures above 250 degrees Fahrenheit because the higher temperatures decrease the digestibility of the food.) We have found that we can put anything in this oven that we'd put into a conventional oven. Of course, there is less space in the solar oven, but with a bit of planning, we can cook a full meal in there. We've cooked squash and pumpkins, potato dishes, pizzas, stews, and soups—you name it—in the solar oven, all with free sunlight for heat.

A fold-out reflective solar cooker (left) and the solar oven.

Most folks still think of solar cookers as some sort of novelty, perhaps a good weekend project for Scouts but not especially useful otherwise. This viewpoint is unfortunate. In part, the skepticism follows from the high cost of prefab solar cookers; some of them cost several hundred dollars! Moreover, many people believe that their yard doesn't get enough sun to make solar cookers practical.

In fact, solar cookers are practical in every region of North America except Alaska for at least six to eight months every year. As for the cost—well, you really can make your own. Here's how.

## Solar "Breadbox" Cooker

If you have a more limited budget or want a "get acquainted" project, you can try making this simple "breadbox" solar cooker, which for the most part requires only scrap or recycled materials.

First, find two cardboard boxes, sized so that one is able to fit into the other, ideally with an inch or two of space all around. If you can't find boxes in the right sizes, you can cut your own boxes from larger pieces of cardboard. Next, cover the inside of the smaller (interior) box with aluminum foil (it is not necessary to cover the inside of the larger, exterior box with foil).

When the little box is placed into the bigger box, the tops of each box should be at the same level. You can support the inner box—so that it is resting off the floor of the bigger box—by placing four small pieces of flat wood or cardboard (we used tuna cans) inside the big box to serve as "legs" supporting the inner box. Once you've placed and glued the legs, pack all the space between the two boxes with crumpled newspapers. Though most people have no problem obtaining old newspapers for the required insulation, you can use many other substances for insulation: cotton rags, straw, dried grass, coconut fibers, whatever is readily available. Though you might be tempted to use those white blown-foam packing "peanuts" for insulation, *don't!* At high temperatures, they may melt or give off undesirable fumes.

Now that you have one box inside another, with their tops level and with the insulation packed between the boxes, you are ready to seal the insulation. This is done simply by taping or gluing pieces of cardboard on top of the opening or "seam" between the two boxes.

The next step is to make a lid for your cooker. If you were lucky enough that the larger cardboard box you found had a tight-fitting lid already, you can use that. Otherwise, you can cut a lid from cardboard. Measure the size of the big box, cut the cardboard at least one inch larger on all sides, and then cut diagonal slots in the four corners and fold down the edges to form flaps so the lid sits snugly on the box. Tape the folded-down corners securely.

Once you have made a tightly fitting lid, you are ready to cut an opening in the lid that is exactly as big as the opening of the inner box. Just cut this opening on three sides of a square, so that you can fold the loosened piece of cardboard upward and create a reflector from the attached flap (see photo, facing page).

The opening in the lid should be covered with a single pane of glass or Plexiglas or a piece of heavy-duty, transparent plastic sheeting. Plastic sheeting is cheaper and easier to install, but glass will retain heat better. We use only glass because it is inert and will not give off fumes. The glass or plastic must be secured to the inside of the lid by glue or silicone caulking. Make certain that the glass is securely mounted before proceeding.

See how the flap that you cut on the lid for the opening can fold up and down like a cover? Line the inside of this flap with aluminum foil, and you have an "automatic" reflector. When the solar cooker is in use, you can prop up the lid with a stick.

Presto! Your solar oven is complete!

Here are a few more pointers on building and using an oven like this. By planning carefully before you begin the actual construction, you will produce a quality cooker with minimal effort. Rather than obediently following the dimensions in someone else's plan, first see what supplies you have at hand. For example, you may have a good pane of glass, in which case you can adjust the cooker's dimensions according to the size and shape of that pane of glass. Or you may find two ideal cardboard boxes, then adjust all the other sizes accordingly.

To cook in the breadbox oven, you can place a black metal cookie tray or a pie pan in the baking compartment of the cooker. To absorb a maximum amount of heat, all cooking pots should be black and should be covered, but if you allow extra time foil will work pretty well. When using regular recipes in a solar cooker, at times you must allow at least twice as much cooking time as needed for a conventional oven.

## Solar Reflector Cooker

We have also built one of Dan Halacy's solar cookers, first described in his 1959 book, *Fun with the Sun*, and in his more recent book, *Cooking with the Sun* (cowritten with Beth Halacy; see the bibliography). Solar cookers work either by absorbing heat within a closed, insulated box or by reflecting the sun's heat to a focused point on a cooking container. Some cookers use both methods. Dan Halacy's cooker uses only reflection.

This reflector cooker utilizes a cardboard disk covered with shiny aluminum foil and propped up facing the sun. A grill mounted on a stand is then placed next to cooker, with a black-bottomed pan or pot (to absorb heat better) resting on the grill. The sun's rays reflect off the shiny surface of the disk, which is curved (concave) to focus the reflected light. The disk can be tilted and turned to focus its beam of concentrated light directly on the cooking container. Because there is no stored heat, the temperature the oven reaches depends on air temperature, clearness of the sky, and focal point, but we've heard of people achieving 300 degrees Fahrenheit or higher with this cooker. With that amount of heat you can bake bread, brew coffee, cook vegetables, fry eggs, boil water, cook soup, sauté *escargot* or mushrooms, and so on. The main drawbacks of the Halacy design are that this cooker is difficult to transport, because of its size and number of pieces, and you should never leave it outside in the rain, because the cardboard will get soft and fall apart.

There are many, many other possible designs for solar cookers. We have made different versions in addition to those mentioned here, and have used others made by other people. We have also seen countless variations described in *The Solar Cooker Review*, the publication of Solar Cookers International, a nonprofit group from Sacramento dedicated to the spread of low-cost solar cookers worldwide.

At least a quarter of the earth's population has trouble finding firewood for cooking daily meals. Their "energy crisis" is a shortage of wood. It is in these Third World countries that inexpensive yet reliable solar cookers can have the greatest impact.

In Africa, about 85 percent of the population cooks over wood fires. In areas where solar cookers have been introduced, along with educational programs to promote their use, significant progress has been made in getting the local people to accept solar options, thus easing pressures on endangered forests and raising people's standard of living with a very small investment.

Yet despite the many advantages of solar cookers, an article in *Solar Cooker Review* (summer 1995) described some of the reasons why people sometimes do not use them. In Zimbabwe, for instance, in an area where solar box cookers had been successfully introduced but then weren't always being used, some people missed the smoke from the fire—an aesthetic and spiritual quality—and the way smoke repels insects. Other people missed the warmth of a fire in their home. Others were fearful that someone might steal their food, or even poison it, if the food were left outside to cook.

## Making a solar hot plate

Here's how you can make one of these cookers yourself. Begin with a 4-foot-square piece of heavy cardboard, about a quarter inch thick. This will be the base plate. Draw a 4-foot-diameter circle on this base. Then draw two perpendicular lines through the center of the circle, dividing it into quarters. These will indicate where the main ribs for the reflector support will go.

Now draw the two main ribs on another piece of cardboard, then cut them out. They will be flat on the bottom, where they sit on the base plate, and curved on top, where they will support the concave reflecting surface. You'll also need twelve half-ribs to complete the framework. Using one of your main ribs, trace and cut six more full-length ribs and then cut each of them in half.

Making slits at the center points of the two main ribs, interlock them like a cross, and secure them to the disk with glue or heavy tape, or both. Then attach the half-ribs to the disk, three to each quarter of the circle. Be sure that their outer edge goes no farther than the edge of the circle. If you need to shorten one of these half-ribs, shorten it from the inside. Secure all twelve of the half-ribs with tape and/or glue.

*Rib pattern for solar hot plate.*

Next, complete the framework by attaching cardboard rectangles to the outer tips of the ribs, securing each one with tape. Be sure that these rectangles stand on the circle you drew, creating a rim like the rim of a pie pan. All sixteen of these rim pieces will each be about the same size, but measure each space and custom-cut the correct size of cardboard accordingly.

Now you can cover the ribbed framework with wedge-shaped pieces of poster board, which will be the base for the reflective aluminum foil; these will also form the curve that is doing the reflecting, so you want a poster board that is strong enough to hold its shape and supple enough to bend easily. Measure each wedge-section, then cut it out of poster board.

You'll need sixteen wedges. First tape or glue eight alternate sections, and then secure the eight remaining pieces.

Next cover the wedges with aluminum foil. Cut out sixteen pieces of aluminum foil, each about the same size as the wedge-shaped pieces of poster board. Try not to wrinkle the foil. Using rubber cement, carefully glue on the sixteen pieces of aluminum foil, shiny side up. Again, try to keep the foil as smooth as possible. Voilà, your cooker is done.

To use the cooker, prop up the reflective disk directly facing the sun. A bit of adjusting will be necessary to get just the right angle to provide the maximum intensity of sunlight. Then move your hand around to find the "hot spot."

Finally, set up a grill securely in place so that your food will be right in that hot spot. The grill itself can be any old campfire or hibachi grill, supported in any number of ways. We used some old electrical conduit to make the stand but have used galvanized piping or wood (be careful—the cooker gets very hot).

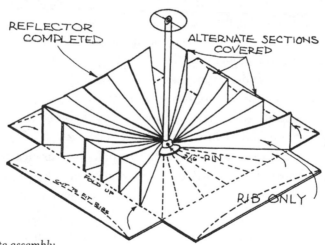

*Hot plate assembly.*
*Ribs support folded up reflector wedges.*

Illustrations from Cooking with the Sun by Beth and Dan Halacy (Morning Sun Press, distributed by Chelsea Green). Used with permission.

In the United States, people generally use solar cookers because they want to, not because they have to. Nevertheless, we enjoy having several solar cookers for use in our back yard. We cook in them mostly in the summer. But even in areas where there is snow, it is possible to make a simple meal with a solar cooker if the day is sunny. We think that everyone should have at least one solar cooker to use when the weather is sunny or in an emergency. In the aftermath of a serious blackout or earthquake, we'd still have the ability to cook meals without power, fuel, or fire.

## FOOD STORAGE

Another area of particular concern when you depend on electricity is the refrigerator. Obviously, if you have a fresh garden, chickens, and an orchard, and know how to find wild edibles, you will be less worried about not having food if the power goes out. Yet even people whose diet is mainly fresh foods tend to need ways of storing food that is not eaten immediately.

Remember that in pre-electricity days, people stored food. Many cultures developed highly sophisticated preservation methods in order to prolong their harvest from farming and hunting. For instance, there are many techniques for drying food, including meats, garden produce, grains, herbs, and fruits. Drying has been one of the primary methods of food preservation for most of recorded history, because it is so simple and requires no electricity or other power. Numerous books explain dehydration options, from simple ways of stringing vegetables with needle and thread for air and sun drying, to the use of more complicated appliances. A number of traditional recipes for drying fruits, vegetables, and herbs can be found in the book *Keeping Food Fresh*, which is entirely devoted to old-time preservation techniques that don't require boiling or freezing. For our part, we use a small electric dryer, or our gas oven, or the sun. The key is to start with foods that are really fresh, then slice them very thin and secure them in airtight containers when they are completely dry.

In addition, people have been "canning" for more than a century. By canning, we mean the heating up of foods in glass jars, which are then sealed with wax or pressurized lids. Food properly canned can be safely stored without refrigeration. The actual storage time can vary depending on what's inside the jar and how you store it.

Another age-old food preservation method is pickling, which uses vinegar as a storage medium and which can be combined with high-temperature canning. As described in chapter 3, we have pickled many types of produce, including garden-grown vegetables and also wild edibles such as seaweed pods (kelp air bladders), radish fruits, and nasturtium seeds. As with canning, there are many way to safely pickle, and many good reference books cover this topic (see the bibliography for a few of our favorites). We generally pickle by soaking the food to be preserved in 100 percent raw apple-cider vinegar. For variety, we have sometimes pickled olives with a very heavy salt solution, 1 cup of table salt to 2 cups of water, in which the food will not spoil even if not refrigerated; food preserved in brine requires repeated rinsing before eating.

In addition to what we produce in our yard and eat fresh and what we preserve, we buy staples in bulk, either as canned goods or dried grains, in order to have a reliable supply on hand. Bulk foods, now available from co-ops or by mail order in every part of the country, are considerably cheaper than single-meal prepared foods, where you are paying a high price for the preparation and packaging. We buy foods in bulk that do not require refrigeration, though a cooler storage temperature does increase their useful life.

Our editor told us about the basement "cold closet" he built, using a design that city dwellers could adapt. His wood-framed box, about the size of two telephone booths, is insulated with "blueboard" rigid insulation. He added a door like a closet door with a flexible foam "gasket" to seal the closure. Then he ran two dryer-vent hoses into the box from above, one that goes down to its floor and the other that vents out through the insulated ceiling. These hoses go out through an exterior wall, drawing in fresh air and circulating out the stale air. The temperature remains a constant 45 degrees Fahrenheit, pretty much year-round, well above freezing and fine for storing bulk grains, root vegetables, and beverages.

We've learned that it is sometimes better to buy smaller containers of food items, rather than big containers. If the power is out and there are just two of you in the household, you may find that you'll open a big can of beans or applesauce (or whatever) but you don't eat it quickly enough, and it spoils. You don't *ever* want to waste food; waste has both physical and spiritual ramifications. During an emergency, wasting food could actually cost you your life.

## Refrigerators

Most Amish people continue to use iceboxes, and in their communities you still find traditional icehouses where ice that has been cut from ponds in winter is stored through the warmer months, insulated with sawdust. Blocks of ice from the icehouse are then placed into the upper compartments of the old, heavily insulated iceboxes to keep foods chilled.

Another Amish technique is to divert water from a nearby spring or stream so that it flows through a cellar room (like a root cellar) where milk and other perishables are kept cold.

While most urban homesteaders don't have any nearby streams that they can divert to keep their produce and dairy goods cold, even for folks in the city there are many viable alternatives to costly and energy-hungry electric refrigerators.

Over the years, we actually purchased three old-fashioned iceboxes at yard sales. We eventually sold these because we realized that we were not likely to actually use them in our daily life and that they would not be the ideal choice in an emergency, because we'd still need to get ice somewhere.

Some modern refrigerators are better insulated than older ones, though the walls of the very old ones are often quite thick. This means that—if you had to—you could convert a discarded modern refrigerator into a sort of icebox, as long as you made sure that melting ice wouldn't drain all over the floor. If ice (or snow) weren't readily available, you could use those "cold-packs" made for coolers, which would help keep food chilled for a while. If you had no power for an extended period, you could take the cold-packs outside during the night so they would cool off for use again the next day.

There are now special refrigerators (for instance, the Sun Frost and Planet brands, both manufactured in California) designed to run on direct current supplied directly by photovoltaic panels. We believe that this is an excellent option, but for those of us who have access to the regular utility grid already, the cost of a solar-powered refrigerator seems prohibitively high.

We're not inclined to disconnect from the grid abruptly but have chosen instead to convert our energy-dependent appliances one by one and to improve the overall efficiency of our daily living.

One such choice was to find the most energy-efficient refrigerator on the market, because the refrigerator that was here in our kitchen when we moved in seemed to be malfunctioning all the time, and sometimes wasn't working at all. The annual handbook mentioned above, *Consumer Guide to Home Energy Savings*, gives detailed comparisons of different brands. In our case, with the help of a friend and neighbor, we researched the question in *Consumer Reports* and learned that the overall most efficient model was a Maytag Plus.

Next, we looked around for ways to get this model cheaper. We located a store that sells "scratched and dented" appliances at reduced prices. When we found two refrigerators, we inspected them and they appeared new, with only minor, merely cosmetic markings. We purchased both, at several hundred dollars off the retail price, with an additional discount for buying two, and paid $25 to have them delivered. One went into our kitchen and one went to the neighbor who had helped us with our research.

We are really happy with this refrigerator. It is quiet and efficient. If there were a blackout, we feel confident that the food inside would stay cold for a long time if we were careful to open the door as infrequently and briefly as possible. Of course, many common foods from the supermarket need no refrigeration, such as honey, bread, or cooking oils. Knowing what does and doesn't require refrigeration is important in allocating limited refrigerator space to the foods that really need to be kept cold.


Our friend Chanel-Patricia had very strong principles, in terms of reducing her dependence on purchased energy. When remodeling her home in Highland Park (a district in the hilly, northeast section of Los

Angeles), she added skylights to provide natural daylight, easy-to-maintain wooden floors, and a large, beautiful stone fireplace. She could actually cook in that fireplace, if she chose to do so.

Her bathroom was also unique. The toilet was in one small room, with the bathtub in an adjacent, slightly larger room. Two people could use the facilities at the same time yet have privacy, a novel but very admirable arrangement.

When she moved in, Chanel discovered that the house had a traditional vegetable cooler, as common in the old days as iceboxes. The cooler appears to be an ordinary kitchen cabinet, but the shelves are made of wire mesh, and at the bottom there is a vent opening into the basement and another opening into the attic. This allows a natural airflow; because hot air rises, the cooler basement air rises and flows over the food on its way to the attic. What a sensible arrangement, all but forgotten by modern builders.

And Chanel chose to have no refrigerator. That's right—she had no refrigerator.

At first we were surprised, because we ourselves store so many frozen foods. Yet because Chanel's research showed her that in many households a refrigerator is the largest user of electricity, and because she was determined to create a home, and create a lifestyle, that was energy efficient, she opted for no refrigerator.

How did she do it?

Chanel utilized most of the time-tested means of food preservation. She purchased many dried foods in bulk, including rice, noodles, soups, fruits and nuts, crackers, and so forth, maintaining a cupboardful of such staples.

She also bought canned goods. At this point, just about any food can be purchased in a can, including condensed milk, soups, pasta sauces, olives, fruit preserves, and vegetables.

Chanel did purchase perishable foods in small amounts at the local market, including fresh vegetables, salad greens, sometimes milk, cheese, kefir, and fruits. She would take care to use up perishables within forty-eight hours. And for storing produce, of course, she utilized her built-in vegetable cooler.

Occasionally when we visited, we noticed that Chanel had perishable items in a dishpan partly filled with water, a wet towel draped over the food. This was her simple form of an evaporative cooler, altogether practical for butter, cheese, and other dairy products.

We always admired what Chanel-Patricia was able to do all by herself. She was constantly finding ways to recycle, ways to save energy, and ways to do more with less.

She passed away from cancer in 1995, and we still miss her. We think of her each time we drive by her little urban homestead in Highland Park.


## WATER HEATING

Where we live, natural gas is the most easily available fuel to heat our water. Even though we know that gas is the least expensive of the various fuels we buy, we still make every effort to conserve and to use our gas appliances carefully.

When cooking, we keep lids on the pots and cook at the lowest possible flame. We have insulated our water heater and all exposed sections of the hot-water pipes to minimize loss of heat.

We are always aware of the fact that gas lines could break whenever we experience one of California's famous earthquakes. For this reason, we have continually explored alternative ways to heat water should the need arise.

We are well aware that Amish families still heat their water on wood stoves. With our wood-burning fireplace, that would be an option for us as well, though not an easy one.

While living in Mexico one summer, Christopher stayed with a family who had a wood-fired water heater. Whenever he wanted to take a shower (there was no bath), he would need to tell them about forty-five minutes beforehand. His hosts would put a few small pieces of wood into the firebox of the water heater, and in a while the water was hot enough for a shower.

Years later, we obtained one of these wood-fueled water heaters. We have taken it to educational programs to demonstrate its simplicity, and we keep it in our garage should we ever need to use it. The model we have is made by the Maga-Mex company. It looks just like a gas water heater, except that there is a small chamber in the bottom where the wood is burned. The incoming line is plumbed to the cold water source, just like a gas water heater, and the outgoing line is plumbed to the hot water delivery pipe. This circulation loop also has a pressure-relief valve, on top of the tank. Christopher's Mexican host family had no other heater, so all the plumbing passed through this water heater, but in a house with a conventional heater this kind of wood-burning unit could be used as a preheater or as a backup in emergencies.

We have used ours in the back yard, connected on the incoming and outgoing sides to a garden hose, which is fine for our purposes because we use this setup only for short periods; for longer-term use, heat would degrade the rubber in a garden hose. Though we have mainly fired it up for demonstrations, we do keep it near an outside bathtub where hot baths can be taken under the stars.

Maga-Mex is not the only company that manufactures these units, but we have discovered an alternative to buying one.

Most people throw away an old gas water heater for one of two reasons: because it leaks, or because the heating element no longer functions. Alas, fewer and fewer people have the desire to fix appliances, so they just buy new ones and toss out the old ones. Even if you could find someone to fix the leak or the heating element, "economics" discourages that frugality, because you can usually buy another water heater for less than the cost of the repair.

As a result, it's not too difficult to find a discarded but nonleaking gas water heater, which can be used as a wood-fired water heater. (Electric heaters are designed differently, so we'd suggest avoiding them.) Remove the sheet-metal shell and the fiberglass insulation, and drain out the cast-iron tank inside. Remove all of the mechanical heating apparatus underneath. Get as much sediment out as you can. Then test the tank to make sure it will hold water.

Remember, the tank must be set on a base of brick or cement—a steel base would be even better. The fire is built right in the bottom section of the tank, in the cavity where the heating element used to be. The hollow channel in the middle will act as a flue, and as the hot flue gases rise, they will help heat the surrounding water.

For the water inlet and outlet, use a galvanized metal pipe. A garden hose just won't do, because the water in the tank gets too hot. Or you can screw in a section of metal water pipe to both the inlet and outlet valves on the tank. The additional sections of pipe are added to serve as extenders to keep the hoses away from the heat. Then with adapters (available at any hardware store) that allow you to go from standard threads to hose threads, attach a garden hose to the pipe on each side.

Generally, you'll find three holes on the top of a gas water heater: one for the water inlet, one for the outlet, and one for the pressure-relief valve. Instead of a pressure-relief valve, you can install a pipe that extends perhaps three feet up. This allows the expanding water room to expand. The problem with this arrangement is that occasionally water may flow up out of the pipe and onto the hot surface of the tank. To avoid this, put an elbow on the top of the relief pipe and add a horizontal extension at least out beyond the water heater so the overflow will not drip onto the heater itself but rather into a bucket.

However you do it, *be sure to install some sort of pressure-relief valve,* because the temperature of the water gets *very* hot with a burning fire underneath it. Steam could build up to very high pressure, and a powerful, scalding explosion could result if you construct a wood-fired heater like we're describing here without having some sort of pressure-relief system.

As noted already, the little compartment where the heating element had been should be sufficiently spacious for a small fire. Start burning some wood there, not letting the flames get too hot until you've tested all of the connections. You can put a little grill in the opening to stop sparks. The cast-iron tank typically has an opening only on one side and otherwise extends to the ground, making this a safe place to make a fire (providing you're doing it on a safe surface), and there is already a built in "smokestack" up through the middle of every gas water heater.

When you turn on the incoming garden hose, pressurized water is forced into the tank and begins to be heated by the fire. If you have no valve on the outlet, then as soon as the tank is full its water will flow out the other hose. We suggest that you put a metal valve on the outlet, even at the testing stage, so you have control of the hot water. Be sure to install a pressure-relief valve, especially if you intend to put the heater indoors (which we don't recommend); explain to one of the salespeople at a hardware store what you are doing and the size of your water heater's connections (typically 1 inch), and they should be able to supply a pressure valve and the appropriate shutoff valves.

Of course, if you find that this kind of water heater works well and you want to move it inside your home, you must install a vent to the outside as sturdy as a wood-stove chimney, much more heat-resistant than a conventional galvanized water-heater wall vent. You don't want to burn down your home in order to save a few bucks. On the other hand, if we were ever to set up our Maga-Mex for ongoing use, we would enclose it in a cinder-block "shed" just outside.

## Solar Water Heating

We have also experimented with solar water heating since the late 1970s.

For instance, once we took a discarded gas water heater, removed its shell, spray-painted the tank black, and set it out in a sunny part of the yard, plumbed so that it could receive a garden hose at the inlet and another hose at the drain. Although you can stand it upright, the heating will be most effective if the tank is tilted at an angle equal to that location's latitude (for instance, at 34 degrees above the equator, the tank should ideally be sitting at a 34-degree angle above horizontal).

It was summertime, and we set the tank up on a 55-gallon drum next to our outside bathtub. When this black tank had been sitting in the sun all day, we were able to drain the hot water (about 100 degrees Fahrenheit) into our outside bathtub for a wonderful dusktime bath.

Had we been out camping, we would have considered this an amazing luxury. However, unless you really enjoy taking baths or showers out in your yard during the day, or unless you have no other choice, bathing in the yard has some drawbacks, including the lack of privacy. And once the sun goes down, that 100-degree water starts to cool off quickly.

So although we enjoyed the ease with which we could obtain hot water from the sun, we were determined to improve on this very basic system so that the water would be kept hot during the night.

## Solar "Batch" or "Breadbox" Water Heater

The next experiment was building a solar water-heating "breadbox," also called a batch heater because it heats a "batch" of stationary water rather than a continuous flow of circulating water. We took that same black-painted water-heater tank and constructed an insulated box around it. This allowed the water to absorb heat during the day and not lose all that heat at night.

It's important to understand a few basic principles. Heat from the sun causes water to expand. Also, the colder water, which is denser, is heavier. As a result, the hotter water in your tank will rise toward the top, and the colder water will sink to the bottom. When additional cold water is introduced into the bottom of a tank, that incoming water will force the hot water out of an exit pipe on top. Of course, some mixing occurs, so you cannot usually take advantage of the hottest water this tank can achieve.

There are many types of batch water heaters. Some have their water tank standing upright and others have the tank lying down. There is actually less mixing of cold and hot water in an upright tank than there is in a tank oriented horizontally. Though upright is better from that standpoint, it can be a challenge to safely support a 400-pound box leaning back at, say, a 45-degree angle—which is the angle that optimizes solar gain at 45 degrees latitude—unless, of course, it is simply "leaning" on your roof. An insulated box around a horizontal tank is easier to construct and has little danger of falling over, but the loss of heat due to hot and cold water mixing is dramatically greater.

Even so, taking all factors into account, and considering our location, skills, pocketbook, and so on, we chose to build a horizontal breadbox back in 1980.

For our version of this solar heater, we experimented with different arrangements, first setting the box up off the ground on two 55-gallon drums. When we wanted to take a bath or shower, we'd simply let the water drain out with the force of gravity. That was practical, to a point, but although the incoming, pressurized cold water was supposed to force the solar-heated water out through the other hose, due to mixing of cold and hot there seemed to be very little truly hot water coming out. Yet many folks in remote areas use a solar water heater just like this, as it will work fine during a sunny day.

To eliminate heat losses at night, you can use an insulating layer (blanket, old sleeping bag, piece of rigid foam insulation, or another door) with which you cover the water heater at night. That means that every night you'll need to cover the glazing and every morning uncover it, which may seem to some people like a great deal of maintenance. And if your batch water heater is up on the roof, you won't want to cover and uncover it every day. A better solution is probably to use a double-paned

## Building a batch solar water heater

For the sake of discussion, we will offer some construction pointers for both horizontal and vertical batch heaters, although we ourselves have built only the horizontal style.

Start by assembling the end pieces. If you are making a horizontal batch heater like ours, you can build a triangular box that will lie flat; start with a 48 x 48-inch piece of plywood (at least ½-inch in thickness, as thinner plywood is too flimsy), and cut it diagonally to make two triangular end pieces. If you choose to build an upright (therefore rectangular) box around the tank, the two side pieces will each measure 24 x 48 inches.

Onto the inside of each end piece, we screw or nail another piece of plywood that has an involute curve cut into it. This is for the purpose of creating a reflector to cast more light onto the water tank (see the accompanying illustration). Then, with 2 x 4s, attach the two end pieces to form the skeleton frame of your box, using standard carpentry techniques. The inside of the box should be at least one foot longer than the tank, in order to leave room for insulation and plumbing attachments. Nail or screw a plywood back and bottom to the end pieces to form the triangular box, or a plywood back to the sides of the rectangular box.

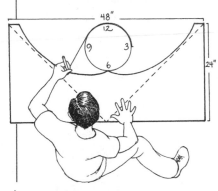

*Drawing the cusp shape onto triangular or rectangular plywood. Stand tank on plywood. Attach a pencil to a string and secure the string to the 3 o'clock position of the tank. The pencil should reach just to the 6 o'clock position. As you unwind the string, the curve is drawn. Reverse for the opposite side.*

Before you insulate, drill the holes you'll need for water pipes. Our triangular box needed six holes: one for the cold-water inlet; one for the hot-water outlet; and four more for the two 1-inch galvanized pipes that will be fastened to the box by two holes in each end; these pipes are needed to support the 400-pound tank within the box, or you could use an angle iron bolted to each end.

Now insulate the inside of your box with rigid foam from the lumberyard, or with some other insulating material. We used old acoustical tiles; they may not ultimately be the best insulator, but we found a free and abundant supply of them. Fasten the insulation to the wood sides on all sides of the tank. If flexible fiberglass batt insulation is used, be sure not to "pack" the insulation in tightly, because insulation's effectiveness depends on the air spaces in the material.

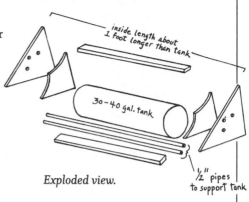

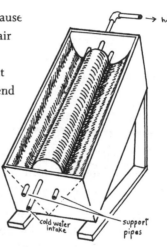

*Exploded view.*

Once the box is insulated, measure out some flexible cardboard and tack it from end piece to end piece for each of the involute curves. This will form the structure for two reflective surfaces. Fill the hollow cavity (beneath the cusp) with fiberglass insulation, recycled styrofoam cups, or egg cartons. Next attach aluminum foil or sheets of reflective Mylar to the entire inside surface of your box. Rubber cement works well, as do spray-on adhesives. All seams should be caulked.

Now you can set your water tank onto the two support pipes, and then attach the plumbing. Once that's done, fill it up and check for leaks. No leaks? You are ready to cover the south side (the side facing the sun) of the box with glazing. Clear plastic will work, but glass is better, for instance, a recycled door; this will be more durable and more effective at retaining heat.

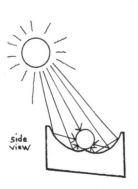

159

(or even a triple-paned) glass door for the front of the heater. On the other hand, the problem with multipaned glazing is that much less heat gets through those insulating layers of glass, resulting in less solar gain.

The most common way of integrating a batch water heater into a home's regular water-heating system is to plumb the solar breadbox so it functions as a preheater for the conventional heater, which in our case is a gas heater. This means that the incoming, unheated water is routed first through the solar water heater then sent on as heated water into the gas or electric water heater, before flowing to a sink or bathtub. When the sun heats the water to a high enough temperature, our gas heater will not kick on, in which case the conventional water heater serves as a storage tank. With a system like this, it's important to insulate the water line running from the solar water heater to the regular water heater to be sure that you don't lose heat along the entire length of the pipe.

Another consideration when plumbing a batch water heater into an existing water system is where to place the breadbox, which is outside the house, in relation to the other water heater, which is often in a basement. Remember, if the breadbox is not really well insulated, the solar-heated water in its tank will cool off at night. Because hot water rises, in such a situation the hot water in the gas or electric water heater can rise and then back through the transfer pipe, which means that you could also be losing the heat from the regular water heater as well—an overall net loss. A better configuration is to situate the breadbox heater *below* your conventional water heater (this usually means not on the roof!). That way as the water in the breadbox heats, the hot water will rise into the conventional water heater. If the water temperature in the gas or electric heater remains sufficiently high, that heater will not switch on. Also, in cases where water might flow "backward" from the house water heater to the solar water heater, you can use one-way check valves that prevent that kind of thermosiphoning.

Our first breadbox solar water heater was built around 1980 for about $78 in out-of-pocket expenses, at a time when most "experts" were saying that you would spend $500–600 to build one. Of course they were talking about going out and buying everything, whereas we were able to use scrap materials and odd supplies that we had lying around. Breadbox solar water heaters are surely the cheapest solar water heaters to build, with the least cost per BTU (British thermal unit, the measurement used for units of heat). This means that the initial investment needed to produce (or save) each BTU is the lowest. For various reasons, though, breadbox heaters are not regarded as the most efficient type of solar water-heating system in terms of operating, heat loss, and maintenance. Even so, if you have a very modest budget, this is an excellent weekend project that will provide you with a firm foundation in most of the principles involved in solar water heating—as well as providing you with immediate savings in the cost of hot water.

### Manufactured Solar Panels

We always wanted a "professional" solar water-heating system, but had found the cost prohibitive for our budget. But then we got to know Michael Butler of the company Solar Performance, which is here in Los Angeles. Butler had attended several of our wilderness walks, and we got to learn about his business. In the past fifteen years, he has designed, installed, and maintained hundreds of solar water-heating systems on Southern California homes. You can imagine our joy when he called us one winter day in 1998 to ask us if we wanted a free solar water-heating panel and the insulated tank that went along with it. A customer from several years ago was having a new roof installed in preparation for selling her home. Rather than take down the solar system and have it put back again when the roof was done, she opted to have Butler "dispose of it." Naturally, we said yes to his offer.

He brought the panel and tank over the next day and gave us instructions for how to install it. We read the instructions carefully and asked many questions, but deep down inside we knew we'd eventually be calling Butler to have him install the system. A whole year later, in 1999, we finally called and arranged for him to do the work.

The installation took two days. On the first day, Butler went into our attic to "beef up" the ceiling rafters. None of us wanted that tank (400-pounds-plus, full of water) to drop through into our bedroom someday. Having measured carefully to find the exact location, Butler placed extra 2 x 4 uprights in the attic directly under the section of roof where the new tank would go. He then built a redwood frame that would hold the tank on the roof and another frame that would hold the solar panel. We carried the empty insulated tank up onto the south-facing slope of the roof and placed it in position, then we carried the panel up and put that in position; and Butler connected the two pieces. The panel was lighter than the tank but more difficult to carry due to its awkward size, about four by eight feet.

The next day, Butler installed all the copper pipe that connected the solar components to the existing water system. The solar heater was plumbed to serve as a preheater for the two gas water heaters in our home. Precise details will vary from house to house, but here's what he did. The incoming cold-water line, instead of going into a gas water heater, is now sent into the solar panel for heating. As the water is heated, it fills the adjacent insulated tank. Then, when we turn on the hot water at our sink, the pressure forces the water into the gas water heater through the system, and ultimately out the tap. As long as the sun has heated the water from the panel and roof tank adequately, the gas heaters do not kick on. Of course, Butler thoroughly insulated the water lines leading from the rooftop solar components to the gas heaters in our kitchen and the garage. And because there is a possibility that the solar-heated water could get *too* hot (which could damage the gas heater), there is a mixing valve with a thermostat located in the pipe just before the solar-heated water enters the gas heater. Should the incoming water be too hot, some cold water will be mixed in to bring the temperature down.

On the first full day that the system was working, we felt the pipes full of the solar-heated water, which were quite hot. Butler smiled broadly as he checked out the arrangement. "You're in the solar age, now!" he bellowed.

In the 1970s and early 1980s, two special factors reinforced a general public interest in solar water-heating systems: the "energy crisis," with its expectation of endlessly rising oil prices, and federal and state tax credits intended to encourage home-scale adoption of solar technologies. According to Butler, when oil prices didn't skyrocket, and when the credits were discontinued in 1985, most of the companies that had been in the business of installing solar water heaters switched to other work or went out of business.

And yet there are still plenty of viable, time-tested commercial solar water-heating systems available. Knowing your needs before you make a purchase will help you choose the best system for your situation.

When you see water-heating systems in mail-order catalogs, don't forget that the cost listed does *not* include installation. You must either install the system yourself or hire someone to do the plumbing and construction work, so don't forget to count installation in your projected cost.

We have frequently seen used solar water heaters advertised in local newspapers. Although we have only talked with two of the sellers, they both had similar stories. One was reroofing in preparation for a house sale and didn't want to pay to put the system back again. The other was selling a condominium and didn't want any complications to hold up the sale. She was the only one in her entire complex with such a system, and there were certain confusions within the association of owners regarding solar heaters. To simplify the sale, she simply had it removed.

Obviously, anytime you choose to produce your own power, there will be obstacles and problems to overcome. This is true for heating your water with the sun, and it's true for using photovoltaics, wind energy, or methane generation. Each approach requires that you educate yourself about the necessary operations and maintenance requirements, and when living in a densely populated urban community, in addition to the technical challenges you may have to face diplomatic or "public relations" problems. People are very ignorant about renewable energy systems, and ignorance sometimes results in suspicion or outright fear.

As such problems attest, "freedom from the grid" is not free. Nevertheless, we think that heating water with solar energy is a good option, and making this choice is good for the long-term health of the planet. While you supposedly cannot actually tell the difference between solar-heated water and gas-heated water, we really feel good that we are now letting sunlight heat our water. To some people that *feeling* may not seem like a tangible benefit (compared, for instance, with saving money), but to us it's a very genuine advantage.

The first time Dolores did a "brush bath" (in a hot shower, vigorously using a brush to slough off dead cells and open the pores of the skin) with our solar-heated water, she had a feeling that she describes as "light and golden." This feeling came not from "saving money," nor from "getting the better of the utility company," and not from being "eco-cool." As Dolores has written, "Maybe what I experienced in my bath was the Feeling of Joy. Until this morning I hadn't been able to put these uplifted moments into more than very general words. We were having a lively, happy conversation about how glad we'd be when the next solar apparatus was installed, the one that would power my study room (lights, computer, fan, etc.). Suddenly I saw, with my 'mind's eye' that the spiritual presences involved in solar energy are 'lighter and cleaner' than those involved when one plugs into the utility company's power. It was these 'better' presences that I had experienced as a nearly ineffable feeling that first time I used our roof-perched solar water heater."

## SOLAR ELECTRICITY

For a long time we wanted to utilize the sun to produce electricity. Photovoltaic (PV) cells, originally a by-product of the aerospace industry, are no longer a space-age pipe dream but a here-and-now viable technology, an example of a high-tech miracle coming home. Specially made silicon wafers or chips are wired together in flat modules, and when exposed to sunlight the free electrons in the cells produce a flow of electricity that can be tapped to power electrical devices and appliances—pumps, lights, computers, audio-visual equipment, motors, any devices powered by electricity from an outlet. However the scale of solar-electric systems varies enormously, from tiny installations that power a few highly efficient devices to huge systems that meet the range of electrical loads typical of a conventional modern building. Because it is possible to spend a lot of money initially on a solar system, you have to do your homework to answer a number of key questions.

Photovoltaic modules produce low-voltage direct current (DC), which is either stored in batteries or (in very simple systems) used to run appliances directly. Do you want to use batteries for storage, or only use power when the sun is shining? Another question is, do you want a 12-volt system, using power in the low-voltage DC form in which it is produced by the PVs and stored by the batteries? For instance, many boats and RVs have 12-volt electrical systems. Or do you want a system in which the DC electricity is "inverted" into higher voltage alternating current (AC), typically 120 volts, the form of power used to run a conventional household? If so, you will lose as much as 10 percent of your solar-generated power through the inverter, a sacrifice most people now seem willing to make for the convenience of using standard appliances and wiring.

Do you want just an emergency backup system, to be switched on to provide minimal "life-support" services if the utility blacks out? Or do you want a hybrid "intertie" system, where electricity produced by solar cells is combined with a regular grid-based electrical hookup? A "net-metering" inverter, which is more expensive, will convert the PVs DC current to AC for your household, but it is also wired into the existing power grid so that when your solar array is producing more power than you are using, your meter will turn backward. This means that you are a net power producer and are *earning* money.

Before going too far in researching and pricing solar components, it is very important to conduct an energy survey of your daily household electrical needs, because every family is a bit different. Needless to say, before spending money on hardware, you will need to educate yourself not only about how renewable energy systems work, but also about your home's energy use patterns and the solar potential (how much sun you have, or "solarization") of your locality. Many good books are available to help in this process, including *The Solar Electric Home* by Joel Davidson, *Practical Photovoltaics* by Richard Komp, *The New Independent Home* by Michael Potts, and *The Solar Living Source Book* by John Schaeffer and the Real Goods staff. When we were exploring our solar options, we also pored over articles in magazines such as *Home Power* and *Mother Earth News*, and *Natural Home* (remember to look for back issues). Another way to get started is to take an introductory class or hire an expert to talk you through what opportunities exist in your situation.

The two reasons photovoltaics are not more widely used today are the initial cost and the propaganda. The cost has steadily dropped. The cost to produce a watt, for example, in the early 2000s is one-tenth of what it was in 1955. But unless you live in a location not served by an existing power line, the renewable energy path continues to seem more expensive in the short term than using fossil-fuel- or nuclear-generated utility power, at least according to the ways most people calculate expenses.

Even so, think about what's really occurring. When you pay a regular electric bill, you are paying as you go, and the regulated power company is allowed to make a profit. In some areas, the monthly energy bills can really mount up. Moreover, in our society the costs of generating energy with fossil and nuclear fuels is also subsidized by public money, giving us a false view of the price. When *all* the costs of creating and transmitting power-plant electricity to your home are calculated, that electricity is arguably more expensive than locally generated solar-produced electricity. When we say "all the costs," we're including the cost to the coal, oil refinery, or uranium workers' health; the health of the local environment where raw materials are mined and burned; the infrastructure used to transmit energy to your home; all the hidden government supports and subsidies; and even the cost of military interventions to maintain access to energy resources.

With solar, you pay the real costs up front, which is admittedly a significant sum of money. And yet for an accurate comparison, figure how much you actually pay to the local power company over the course of five years using fossil fuels and supporting their system. For a stand-alone PV system, you pay for solar modules, batteries, an inverter, a charge controller, and the various wiring, junction boxes, and conduits, some of which are standard components of any electrical system. It may then take several years for an average PV system to earn back in savings the extra expenses beyond what utility power would have cost over the same period. From that point on, however, your power is "free"—paid-for in advance, and protected from market fluctuations and insecure supplies. Of course, as an independent power producer, you will need to maintain your system, but beyond adding distilled water to batteries periodically, household solar systems require very little maintenance on a daily basis. Ongoing long-term expenses include replacement of batteries every six to ten years (less frequently with more expensive batteries), but the solar modules themselves may be good for many decades of power production.

In our case, we have long been interested in solar because we believe it is *right* to support renewable energy sources. Although many analysts of the solar industry predict that there will be no great reduction in cost or great rises in efficiency in the near future, a PV system is already a good option today if you plan carefully and do your homework.

We began our involvement in solar electricity at least twenty years ago by buying small gadgets run by the sun. We have several solar devices that recharge our nickel-cadmium (nicad) batteries. Over the years, we have purchased solar-powered radios, lanterns, lamps, and walkway lights. We also have a small solar module that sits on the dashboard of the car and plugs into the cigarette lighter, producing a trickle or "float" charge to keep the car battery fully charged. When we used to go to a monthly flea market to sell things, we would have our truck parked right next to us, and with this little PV plugged in, we'd be able to play the car radio or cassette player all day without wearing down the battery.

Beyond incidental electrical devices, we have also wanted to use the sun to produce a greater share of the electricity we use for household appliances on a daily basis. We admit, however, that we were as ignorant as the general public in not knowing how to utilize photovoltaics for more extensive energy production. We had even been invited to buy PVs "on sale" and "wholesale" many times, but held off making any purchase because we weren't sure what else we needed in order to do the complete job properly and safely. As we've learned with nearly every project we've undertaken, if you study and do research, you are more likely to find a way to get just what you need at the best price.

Finally, in early 1999, with Y2K concerns mounting, we decided it was the right time to plunge in and learn what we needed to know. We began by studying various books on the subject. We also paid a visit to our friend Ted Baumgart of La Crescenta, north of Los Angeles. Baumgart has been concerned about self-reliant living his entire life. His backyard pool and a fishpond waterfall are powered by PVs, as is an electric train that runs throughout the back yard. He has been remodeling—rebuilding, really—his entire home so that it will be more energy efficient. This work is ongoing, and eventually he plans to have sixteen PV panels on the roof, with an inverter tied into the grid—which will run his utility meter backward if he produces more power than he uses—as well as some batteries for emergency backup power. His home is part of an annual national tour of homes that utilize solar power.

We got a lot of practical advice from Baumgart about putting a PV system together, beginning with advice we had already heard many times: start designing your solar system by doing an energy survey. This involves listing all the electrical devices you plan to use, how many hours each day you will use them, and how much power each one takes. If you plan to create a system to power your entire home, you need to do the energy survey with all of your household's energy loads. If you are beginning with a solar system that supplies power to one part of the house, for instance, your office, you can do the survey with the power usage of only that room. We read an article about some people in Indiana who had an off-the-grid room, where all the electrical loads were supplied by solar—along with a TV powered by an exercise bicycle (if you stopped pedaling, the screen went dark). *The Solar Living Source Book*, mentioned earlier, includes an energy-survey form that you can photocopy and fill out, in addition to wire and conduit sizing charts, battery-wiring diagrams, wattage listings for common appliances, and many other useful resources.

CHRISTOPHER NYERGES

*Wade Webb installs photovoltaic modules on our roof. Behind him is our solar water heater.*

Completing the survey, you arrive at a figure you can use to determine your average daily, weekly, or monthly usage. One of the other benefits of doing an energy survey is that this process forces you to look at all the ways you are using electricity, including the very wasteful "phantom loads"—devices that are on all the time, including refrigerators, electric clocks, answering machines, automatic coffee machines, computers (unless you switch them off), televisions with remote controls, and so on.

It is worth reflecting on the way most people in the world live today, and worth realizing that we don't *need* electricity to live. With that realization in mind, consider the fact that if you were to cut your electrical usage in half, you could create a significantly smaller and less expensive PV system to handle your needs.

After much discussion, we purchased a 1,000-watt inverter (which can continuously supply 800 watts of AC), two 98-amp batteries (rated for 250 cycles, meaning they will last from six to eight years), two 64-watt solar modules (which should last twenty-five years or longer, perhaps far longer), and all the necessary hardware from Solar Webb, Inc., in Arcadia, California (www.solarwebb.com). We also hired the company's owner, Wade Webb, to install the system. Webb has been a practicing electrician for about twenty years and has been doing solar installations exclusively for the past six years, so we figured that it would be best to have him set us up while we watched and learned.

Webb studied our roof to analyze its solar exposure or angle, and determined where to put the two PVs. He carefully wired the two modules together in series, then he mounted brackets to the underside of each unit and screwed them into the roof. To prevent leaking, he caulked the holes in the roofing with silicone before putting in the screws. Then he threaded the "hot" (positive), ground, and negative wires through a junction box, down through a safety shutoff, and into the battery box, which is a waterproof Rubbermaid container kept outside. The two batteries are connected in parallel, wired to a charge controller that is kept near the batteries and that coordinates the system's interactions, preventing a battery overcharge and shutting down the inverter if the batteries are drawn down too low. All the components are wired into that controller—the batteries, the wires from the panels, and the wire leading to the load side, meaning in our case, to the inverter. We mounted the inverter just inside Dolores's office, and Webb had to drill a hole through our stucco wall to connect the inverter to the controller, which is outside. Webb drilled two holes through the side of the Rubbermaid box for the lines coming in and out. The last step was to connect the batteries.

The entire installation took Webb about five hours. We kept the inverter turned off until the following day so the batteries would be fully charged. What an exciting moment for us: to turn on the inverter, see that our batteries were charged, and then plug in the power strip for the computer, printer, lamp, and phone machine. We started the computer and it came on just as it always does, but this time it was off-the-grid. This made us feel good. We both find that it's somehow cleaner and purer to power our devices this way. This feeling seems too subtle for explanation.

This moment reminded us of the day when we had our solar water-heating system installed, when we had laughed with Mike Butler who installed the system about whether solar-heated water would feel better than gas-heated water. We felt that this was something that is good and right to do, intrinsically "better" than relying on increasingly scarce fuels.

Dolores went about her day's work, but we kept "inspecting" the inverter and then going outside and "inspecting" the PVs. I suppose that

## Analysis of Our Solar Electric System

- *Two Uni-Solar 64-watt panels, wired in series. This is the production department of the system. While their nominal output is 128 watts at peak production, the actual CEC rating is 60 watts. This difference is the difference between the theoretical and the actual wattage you can expect.*
- *Two Deka 12-volt lead-acid batteries, rated at 98 amp hours, connected in parallel. This is the storage department of the system, from which the home draws direct current (DC) power.*
- *One Portawattz 1000, a 1,000-watt inverter that converts the 12-volt DC to standard household power, which is 110 volts alternating current (AC). The inverter is "de-rated" between 5 to 10%, meaning that with a 1,000-watt inverter, only about 800 watts are usable.*
- *One SunSaver-10 charge controller, which allows up to 10 amps of PV power into it. Because our system's two 64-watt solar modules are rated at 3.88 amps, we will need to upgrade the controller when we install more PVs.*
- *The appropriately sized wires, junction boxes, and connections as determined by the National Electrical Code Article 690, in the code guidelines on photovoltaics. Any licensed electrician should carry this handbook, and the photovoltaics guidelines are included in their entirety in The Real Goods Solar Living Source Book.*

we were like children, excited that we were finally doing what we'd wanted to do for a long time.

Our goal with this PV system is to have a power source to run one office completely independently—computer, printer, phone machine, lights, fan, and perhaps some other devices. In the event of a blackout (a storm, an earthquake, too many birds sitting on an electrical wire in Northern California . . .), we would continue to have power for about three days from the batteries alone. And as long as we had sun, we'd be able to keep the batteries charged so as to have a few lights and maybe power a small television in addition to the computer.

We had originally planned to have our solar system power an office using a laptop computer, a light, a fan, and a small cooler. We decided,

---

*Usage Analysis: An Example*

Our desktop computer uses 150 watts to operate. If we run the computer eight hours per day, that is 150 x 8 = 1,200 watt-hours per day.

Our two solar panels each generate 64 watts with approximately 5 hours of usable sun per day, which means (64 x 2 = 128 watts) x 5 hours = 640 watt-hours generated.

Therefore, though this system would be fine to operate a laptop computer, some lights, a fan, and a telephone machine, it provides only half the power we need to run our desktop computer system.

You can do this kind of analysis with each of your regularly used appliances by looking in the owner's manual—or "specifications" summary, always found on the unit itself—to find out the wattage required for operation. Remember this equation

amps x volts = watts.

So if you only know the amperage, you can multiply that by 12 (in a 12-volt DC system) or 120 (in a more typical household AC system) to learn the wattage.

however, to run a desktop computer on the system, a lamp, and the phone machines. We soon discovered that although our PVs can power these loads for a few hours, we then begin to run out of power. So, as soon as possible we plan to add a few more PVs to increase the amount of electricity we're generating and another battery to increase our storage capacity. Currently, our PV system regularly charges Dolores's halogen lamp, cell phone, digital camera, and freeplay radio and flashlight.

Recently we added a 12-volt attic fan that is operated directly by a small solar module. This small fan maintains a continuous airflow through our attic during the heat of the day, drawing heat upward and keeping the whole house cooler in the summer, which eliminates the need to ever turn on the energy-guzzling air conditioner. We purchased the attic fan from Wade Webb, and the PV we got out of the trash. That's right! At one of the Preparedness Shows in Sacramento, California, a vendor threw away a solar module because he had dropped it and cracked the surface. We took it out of the trash, had it tested, and were delighted to see that it produced a full charge. But you can't put a panel with a cracked surface on your roof, because moisture getting in will quickly end its useful life. So we purchased a sheet of plexiglass from a glass shop, and mounted this to the surface, sealing it with clear silicone. Now we have a solar-powered fan keeping the house cooler in summer, at half the cost that a complete new set of components would have cost, and at a fraction of the cost of buying and operating an air conditioner. Truer words were never said: one person's trash is another person's treasure.

Eventually, we would like to install a net-meter inverter. We would stay on the grid but would aim to produce much more of our own power. We love the idea of actually turning the meter backward whenever we are producing more electricity than we are using.

These days, we often think about the sun with amazement and gratitude. There it shines, providing what is required for all life, in the garden and in the home and on the earth, with no harmful environmental by-products or side effects—and yet countless "experts" continue to insist that the sun is not yet a viable source of power.

In our home we are committed to being part of the solution. We really feel an inner gladness to be now among the energy producers, no longer just consumers.

# 8 Trash and Recycling

*We must become increasingly aware of our ever-present tendency to use the mercy of a loving God, and His readiness to forgive, as an excuse for careless living.*

—FRED RENICH, *WHEN THE CHISEL HITS THE ROCK*

As longtime "practical recyclers," we've become far more attentive to each and every bit of "trash" that previously we might have discarded. We've found that most items that we once we would have tossed out are valuable resources that can be recycled or reused. And since we've started taking recyclables to a local public recycling center, we've felt good about not only being part of the solution, but getting paid besides.

For instance, we spent an hour one morning driving through our neighborhood on trash pickup day collecting old newspapers. We told a friend of ours, James, how glad we were to get $5 for a load of papers.

James (the skeptic) wasn't impressed. "An hour for five dollars! That's a waste of time," he sniffed.

"Now just a minute," we quickly responded. We told him that we feel good recycling, not merely for the money but because we're literally saving trees. The money is just the frosting on the cake. "Anyway," we told him, "the five dollars we received for the papers was just a fraction of our monetary reward."

We explained that while we were collecting the papers, we also salvaged two perfectly good picture frames. These were wooden, about three feet by four feet. And we found a 5-gallon wooden plant pot. "The dollar value of these items, conservatively, is fifteen dollars," we told James.

"Okay, so you got a few knickknacks on the side," puffed James.

"Wait, there's more," we responded.

We told James how Dolores is a devoted "coupon cougar." Before we dropped off our 250 pounds of newspaper at the recycling center, we removed all the inserts containing the manufacturers' coupons. It took ten minutes to pull out these coupon pages, and another thirty minutes to clip all the coupons for products that we actually use.

"Oh, boy," snorted James, "so you get a few dollars off at the supermarket with your coupons. So what?"

"Not just a few dollars!" we eagerly told James. We showed him Dolores's system for filing coupons (described in chapter 9). From that one batch of coupons, once doubled—and we go only to the stores that double coupons—we'll save over $76 buying products that we'd buy anyway.

"Seventy-six dollars!" James didn't believe it. He hovered closer to see how much money we'd save on which products. He saw that many of the coupons were not just for 5 or 10 cents, but 40, 50, even 75 cents off. And when doubled, the savings really add up.

"Now, that's not all," we declared. We took James out to the back yard to show him seven 15-gallon planting pots that would cost about $5.50 each at a nursery.

*A view of the tire stairway on the hillside wild garden.*

DOLORES LYNN NYERGES

"We saw these in the plastic section of the recycling center. The man told us that a customer had left them, but they couldn't be recycled. We asked the man if we could take the pots, and he was glad to get rid of them."

James was silent for a moment.

"Hmmm, let's see. . . ." He did some figuring out loud. "Five dollars for the newspaper, $15 value of the frames and wooden pot, $76 in coupons, and $38.50 for the big plastic pots. That's . . . uh . . . wow! That's about $134 and some change! Why, that's not bad at all for a little under two hours of work."

We agreed. The practical and monetary benefits of recycling and "living better with less" are not readily apparent to everyone, but with a bit of explaining even the most hardened cynic like James can come to see their value.

Still, the money aside, we've found a great spiritual benefit in our practice of "living lightly on the earth." Our eyes are opened to the vastness of wealth all around, and to the unbelievable horrors of waste that is the norm here. We shudder to think of the horrible karma that the extravagant wasters are earning for themselves in the future!

By "living lightly on the earth," we participate in the solution to not only the resource crisis facing humankind, but to the problem of greed and careless waste. And *that* is vastly more important than getting $5 for a load of newspaper. (The city recently implemented a curbside recycling program, so we no longer collect recyclables to sell.)

Let's look at some specific recycling ideas. What follows is a combination of projects we have done ourselves and unique approaches we've seen other people use. Needless to say, just as with gardening, carpentry, or saving energy, the more organized you are, the more efficient and therefore beneficial your recycling activities will be.


GLASS

Superheated sand, a marvel to behold—glass is a valuable resource. Consider how difficult it would be to manufacture glass if all you had was miles of beach and a large fire. Granted, the art of glass manufacturing is actually considered primitive—some ancient peoples manufactured glass even though they lacked other technologies—but most of us would never be likely to stumble across the secrets of glassmaking. Yes, this is a marvelous substance.

### Bottles and Jars

The most common forms of glass receptacles we encounter daily are bottles and jars, which many people casually toss into a trashcan as if they were nothing but garbage. It's true that such glass is ubiquitous. Although beverages are increasingly sold in aluminum and plastic rather than glass, and many of the products (such as shampoos) that were almost always in glass containers are now in plastic, glass is still everywhere.

Whether clear, brown, green, or some other color, nearly all kinds of glass can be easily recycled. The simplest way to prepare glass for recycling is to clean the containers right away after use, then sort them into temporary storage containers. In most communities trash hauling and recycling takes place on a designated schedule, either through curbside pickup or by drop-off at a transfer station. Different rules apply in different locales with respect to how much advance sorting the homeowner must do, so ask your neighbors or get guidelines from the local solid waste office; in many cities now all the recyclables can go together into one bin, which is provided free by the recycling program. When recycling day rolls around, you can put out your recyclables to be collected or take them to the local recycling center. In some areas beverage bottles can be returned directly to merchants to "redeem" the returnable containers for money.

But in addition to the option of giving jars and bottles to a recycling center, here are some other practical ways to reuse those common glass containers.

STORAGE. Glass jars are excellent in the garage for storing nails and tacks and screws and drill bits and nuts and bolts and staples and hinges and all those little items that need to stay organized and visible.

One of the more ingenious systems we've seen involved securing the lids of jars onto the undersides of shelves with two wood screws; then the jars could be screwed up into their lids, leaving them hanging with their contents visible and also leaving the part of the shelf space available for other storage. We saw this in Timothy Hall's workshop, using old salsa jars that were all the same size and style, situated at approximately eye level.

Glass jars are also ideal for kitchen storage needs, especially for people who purchase beans, grains, and pastas in bulk, because these staples are best taken out of their paper or cloth containers, where they would be susceptible to spoilage or roaches and rodents. Most sizes are useful—gallon jars for flour, cereals, and so forth, and smaller jars for herbs, spices, loose teas, and condiments.

Solar Tea. We find it humorous that someone is actually making money selling glass "solar tea makers." Many people we know have just taken old gallon jars, rinsed them out, added water and tea leaves, and set them in the sun. We do the same ourselves; we would never consider paying $5 for a jar with a plastic spigot and the words SOLAR TEA printed on the side.

Any glass jar can be used for making solar tea. It can be clear, green, blue, any color. Just set the jar with water and tea leaves or teabags in a sunny location. On a sunny day, a few hours in the sun is sufficient for making warm tea. If you want it cold, simply pour it over ice cubes in a drinking glass.

Candle molds. Virtually any type of bottle or jar can be used as a candle mold, depending upon the intended shape and application of the finished candle. Larger jars, such as those from mayonnaise or pickles, will make good tabletop decorator candles. Narrower shapes, such as those from catsup bottles, spice jars, or wine bottles, will result in a taller, narrower candle.

Candle making is quite simple. First hang a wick (bought from a craft shop) down into the container, securing it on a stick or pencil so that it stays in the middle of the opening. As for wax, we've found that beeswax works best, but there are so many options we recommend that you experiment. Heat up the wax in a saucepan, then pour it into the mold a little at a time, preferably in one-inch increments added no sooner than one hour apart. If you pour all the wax at once into a bottle mold, you're likely to end up with several large airholes once the wax has dried and settled. Be careful: wax can catch on fire if it gets too hot.

Wait twenty-four hours for the candle to harden. Then, using a hammer or similar tool, carefully break off the glass. Of course, all that broken glass can still be collected and added to your recycled glass bin, but be careful with the sharp edges! (See below for an even simpler way to make candles using cardboard juice containers.)

DRINKING GLASSES. If you have a means of cutting glass, it's possible to make many of your drinking glasses and several other useful items. Randy Denham of Pasadena often complained that he could never find good, thick drinking glasses—the type that could fall to the floor without breaking. Failing to find the quality he was after in any stores, he decided to make his own.

Randy observed that certain "disposable" containers—for instance, bottles for alcoholic beverages such as beer, champagne, or liqueurs—are made of a much higher-quality glass than any glassware he had found. So he collected several suitable bottles that he felt would work well, then filled the bottles with old motor oil (any oil would work) to the level where he wanted the glass to break. He heated an iron rod to white-hot (he used a torch to heat the rod, but we've also seen it done on a stove or fire). The rod had to be small enough to fit into the bottle, and he had to fashion a handle for the rod. Once the rod was white hot, he carefully inserted it into the bottle and into the oil. He'd leave that rod in the jar for less than a minute. If everything was done right, he'd hear a "ping." That meant that the glass had cracked, and he would remove the rod, carefully empty out the oil, then gently tap the bottle until the top part and the bottom part separated. If the glass was exceptionally thick, a clean break would be unlikely, whereas with thinner glass, a clean break could be expected. Randy would take the bottom half of the bottle—his new drinking glass—and buff down the sharp edges on his buffing wheel.

Perhaps this seems like a great deal of work to go through for a quality drinking glass, but Randy never thought so. Not only did he convert what other people called "trash" into useful objects, but he loved to show off those beautiful, one of a kind glasses that he had fabricated.

They were special, indeed, for his handiwork had gone into them. Here was a genuine example of the alchemy of recycling, the turning of trash into treasure.

BOTTLE ARCHITECTURE. A unique house can be made by mortaring bottles horizontally into earthen or concrete walls. Typically the bottles used in this way are colored, and they should be capped or corked to keep out dirt and insects. Bottles are usually mortared into walls during initial construction, though sometimes a hole can be made later and remortared, once the bottle is placed. The effect of sunlight shining through the glass is especially lovely, so consider the seasonal solar angles as you choose locations for this unusual form of "window." There are not many complete houses of this kind in existence, but it's not uncommon to find several bottles in walls of older homes in the southwestern United States. In his book *The Sauna*, Rob Roy describes a technique for alternating bottles with logs when building a cordwood masonry wall.

Though we have never actually done this, we have seen some beautiful walls constructed with bottles.

## METALS

Most metals can be recycled, and considering the longevity and durability of these materials, we should find a way to reuse metals rather than simply discard them in a trashcan.

Some recycling centers will accept "tin" or steel, all will accept aluminum, and many will accept brass and copper. All aluminum beverage cans can be recycled, and many cities (such as Los Angeles) collect cans at curbside. It's very easy to participate in a recycling program. Simply wash out the can when you're done with it, let it drain, and then put it in a bag or bin (preferably outside) until it's picked up or you take it to the recycling center (or in some areas back to the store where you bought it).

Often old appliances are full of recyclable metals and shouldn't just be tossed away. You would probably be surprised to discover how much usable hardware can be found on the average tossed-out television, refrigerator, water heater, or dishwasher. For example, before we removed the automatic dishwasher from our home, we unscrewed every part that could be removed. We saved some of the screws and bolts we retrieved in our "miscellaneous" bottle and the rest of the assorted metal was put out for the city to recycle.

There are so many ways to use old cans—the list seems infinite. Some of the ideas that follow may seem obvious, but we live in such a throw-away society that it's worth pausing to reflect before tossing yet another item onto the ever increasing mound of garbage that will be our legacy. The value of using recycled materials such as metal cans is that they are free, ubiquitous, and often very sturdy. And eventually after reuse they can be thrown out (that is, the metal recycled), in cases where that is easier (for instance, recycled-can mess kits blackened by camping) than cleaning them back to a pristine state.

The few ideas given here should help stimulate your thinking.

PLANT FOOD. With recycled cans you can make your own "iron water" for fertilizing potted plants and trees. Ellen Hall of WTI in Los Angeles first showed us how to make this useful garden amendment back in 1976. We fill a 20-gallon bucket with crushed food cans, then fill the bucket with water. Within a few weeks, the cans rust and the water turns brown. That rusty brown water is actually a good source of iron for your plants, and this is an easy way to recycle cans and manufacture a good homemade plant food.

To use the iron water, just dip into the big bucket with a large can and pour the rusty liquid on your plants as needed, watering them as frequently as you would normally.

It is surprising how quickly the oxidizing process occurs. We've seen the entire contents of a bucket rust away to nothing in about seven months. Of course, you can use a smaller or bigger container, depending on what's convenient, and you can continually add crushed cans as the other cans rust away. One concern is that you don't want to breed mosquitoes, so you'll need to keep the iron-water container covered. If the surface of the water ever does become home to mosquito larvae, empty all of the water onto your various plants and refill the bucket with fresh water.

WATERING CAN. A simple watering can is easy to make with a large fruit juice or tomato sauce can. Take a tenpenny nail, punch holes in the bottom of the can, then take wire from an old clothes hanger and attach a handle to the top of the can.

Dip the watering can into a bucket of rainwater or iron water, or fill with tap water, then hold it over your plants. The water just drips out; the intensity of the flow depends on how many holes you punched in the bottom of the can, and with very few holes the drip is quite gentle, ideal for watering sprouts and other plants that are fragile.

PLANTERS. Virtually any used can is a potential planter. The larger ones seem to be better, evidently because the soil dries out more quickly in smaller containers. Even so, some plants actually prefer dry soil; for example, small cat food and tuna cans are good for small succulents and cacti. Larger plants can be put in soup cans, ham cans, or those large cans used for cafeteria food—all the way up to the 5-gallon cans used to ship nuts and other goods in bulk.

Due to the speed with which these metal pots rust away, you should expect to keep your plants in them for only about three years at most, but this is fine, because often at that point you can transfer the maturing plant directly into the ground, pot and all. If by then the can isn't yet in an advanced stage of rusting, punch several holes in the bottom so the roots can freely expand. Planting seedlings directly into the ground this way provides a certain degree of protection against moles and gophers. The main drawback is that sometimes leaving a plant in a can will stunt its growth.

SOLAR WATER HEATER. This topic is covered in more detail in chapter 7, but here is another way to heat small batches of water using inexpensive recycled metal containers. Take a 1-gallon can, paint the outside black, and set the can in direct sunlight. Depending on air temperature and time of year, you'll have hot water in anywhere from a half hour to a few hours. To help the heater gain heat in cooler weather, cover the can with a sheet of clear plastic and secure the edges of the plastic with pebbles or soil.

This is sometimes referred to as the "hobo solar water heater" because of the ease with which a traveler can assemble it.

Heating water in a clean can doesn't present the same problems as heating your water in a plastic bag, in which case the heated plastic may release noxious chemicals into the water. With a can, there should be no problem using that heated water for tea, soup, or coffee, as well as for bathing, washing dishes, or washing clothes.

Tool caddy. Every gardener knows how easy it is to forget your hand tools in your garden. One way to keep tools from getting mislaid on the ground is to hang a large coffee can on the back side of a gate or on a post. You can temporarily store your tools in the can, with a few holes punched in the bottom to allow rain and sprinkler water to drain off. When you're done digging and planting, you can unhook the can and carry the tools back to a more protected storage place.

Camping gear. Campers and hikers fabricate many useful items from old cans. A large can (a gallon or larger) will serve well as a stove. With the lid cut off, you can simply invert the can, punch a few holes toward the top for smoke to escape, and build a fire of scrap paper and twigs in order to cook on the hot upper surface. You can cut a little door on one side of the can to facilitate the feeding of the fire.

You can also use a large can as a stove with the open end up, which in dry conditions is safer, because the fire will be burning on the metal bottom of the can. You can cut a little "door" near the base for the introduction of small firewood, or you can just drop wood in from the open top. If you choose not to cut a small door in the bottom, at least punch a few holes toward the bottom to provide an airflow, because a fire needs oxygen. Larger pots can be rested right on the top of the open can, but smaller pots will need some kind of support such as a stiff piece of screen or grill.

Of course, a recycled can will also work well as a pot. Just add soup or stew and simmer over the fire. A smaller can may be used as a cup or bowl.

All of this might sound ridiculous if you've got a top-of-the-line mess kit and the latest in high-tech camping gear. However, we've used cans these ways over the years and found them perfectly serviceable, and on a few occasions when we hadn't actually expected to be "camping out," we had to improvise our mess kit and were glad to have cans for a backup. When our trip was over, we recycled the can-cookware.

LANTERNS. We've also made a couple of dozen "hobo lanterns" over the years, and find them remarkably useful.

Cut off both ends of a can. A soup-size can is all that's needed, though you could use a somewhat larger can. Run a clothes-hanger wire through the two open ends to serve as your handle, allowing the can to be hung horizontally. Next—and this is the only hard part—cut a small hole in the side of the can (the middle of the "bottom" when the can is horizontal). Heavy-duty tin snips work well, but any sharp implement that can pierce through metal will do. Just be careful not to cut yourself! Once you've punctured the side of the can, carefully expand the hole until it is large enough to hold the base of a candle.

Insert the candle into the hole from below, light it, and the hobo lantern is ready. By bending or curving the wire, you can shape the handle to allow you to easily direct the light where you want it to shine.

*An aluminum can lantern hanging from our back yard apple tree.*

Watch the candle so that it doesn't burn too low and fall out. You have to periodically push up the candle as it burns down.

We've used this kind lantern out in the desert, and while working in the dark in the basement when we didn't have a flashlight. In addition to serving as a good reflector, the can also provides wind protection for the candle flame.

Another hobo lantern involves cutting an H along the side of a beverage can (see photo, above). Open the flaps and insert the candle. Hang the lantern by the pop-tab.

WIND CHIMES. The lids of cans can be saved and used to make wind chimes. It was Ellen Hall who first showed this idea to us. Ellen used lids of all different sizes, small to very large, and they had turned brown with rust, which gave the chimes a rustic appearance.

Using a nail, she made a small hole near the outer edge of each lid and then hung many from a piece of wood. These lids were not like bells when the wind blew; the sound was much more delicate, somewhat "tinny" and tinkly, but pleasant.

These chimes are easy to make and last for years. If one were to take a little time and care to make them beautiful, they would earn some income at yard sales and flea markets.

Here are some more projects for your other recyclable metals.

BIRD FEEDERS. The easiest birdbath or feeder imaginable can be made by hanging an old hubcap from three wires, then filling it with water or birdseed. An old pot lid can also be used in this way.

BICYCLE RIMS AND TIRES. Old bicycle rims are occasionally discarded when they are too damaged to be repaired or merely because the bike owner wants something different. Bike shop mechanics are used to seeing quite a few of these tossed away.

If you're a tinkerer, you'll discover that functional bike rims can be adapted to many interesting projects and experiments. For example, we've seen homemade garden carts built with two bicycle tires. Or the bike rims can be suspended horizontally and used to hold up clothes-drying lines or dog run lines, or in other applications where a pulley mechanism is desirable, with the rope running around the inner track of the rim. Hanging bike rims can also be used to hold up the can-lid wind chimes described earlier.

Perhaps the easiest way to recycle bicycle rims is to mount them on the side of a wall or on a fence, then use them as a trellis for climbing plants. We once saw the entire side of a barn covered with various-size wheel rims, some just touching and some overlapping. They had been secured with hooks, and plants such as beans, ivy, and peas were sprawling up over the matrix of rims. These had been in the outdoors for several years, and they had turned a rusty shade of brown that we found quite attractive. We've since discovered that metal isn't the best material to use for trellises, because it can get extremely hot or cold (depending on the weather), causing the plants more stress than necessary.

We are certain that there must be many more good ways to reuse bicycle tires than we have thought of. We would love to hear your ideas; send us a picture of your ingenious bicycle-wheel project.

WINDOW SCREENING. If you are a gardener, you'll always have a use for an old window screen. Nearly every time we plant a tree, we first line the hole with a piece of recycled screening, which protects the young trees from being chewed to death by burrowing rodents. This is a wise precaution even if you believe you haven't got any moles, gophers, and the like in your area. We've lost many trees and other plants because we believed that there were none of these underground burrowers in our garden, but once you start planting and watering new plants, somehow they know and will gravitate to your yard.

The screen material will slowly rust away, and that rust will add iron to the soil. By the time the tree is old and large enough to need more root space, the screening will be easy to penetrate, and the tree will no longer be as vulnerable to rodents.

If you have a more determined and destructive mole or gopher population, you will need to resort to more serious measures to protect the plants. We've seen chard plants "disappear" into holes in the ground, and likewise carrots and potato plants. To protect your crops, consider building a raised bed with the entire bottom screened across. In a raised bed like this you can grow all of your root vegetables or any other susceptible crops. For instance, asparagus and rhubarb often need protection when newly planted.

METAL CLOTHES HANGERS. These can be lifesavers, if you have them around when you need them. Who hasn't used one to get into a car when the keys were locked inside? Metal hangers are pliable but strong, relatively easy to shape, and can be fabricated into plant hangers, bucket handles, a short-term car radio antenna, or even bicycle toe clips. Ideal for makeshift repairs (for example, holding up a car exhaust pipe or muffler, or for other emergency bicycle repairs), this formerly ubiquitous household item has countless other uses. Yet with the increasing proliferation of plastic versions, metal clothes hangers may soon become obsolete, a thing of the past.

TWIST TIES. The paper-covered wire twist ties used to close supermarket produce bags can be put to any number of uses. Most obviously, you can use them to close up recycled plastic bags. If you raise grapes, vines, and peas or if you espalier plants on a wall or trellis, you can use twist ties to secure stems and branches. For garden applications, the twist ties should be kept with your garden supplies.

Inside, we've used them as pipe cleaners, or temporary fasteners (including for hair). We save them in a kitchen drawer so that they will be on hand when needed. You may not know what you'll do with one now, but there are jobs where nothing else will quite do.

WATER HEATER TANKS. Whether gas or electric, these units are commonly tossed out when they no longer work. More often than not, the tanks still hold water; it's the heating element that usually goes bad. And although it's a relatively simple task to repair water heaters, virtually no one in Los Angeles (nor probably in any major U.S. city) does so, as it is not considered economical to repair a water heater when you can buy a new one on sale for as low as $70. This means that these excellent 30- or 40-gallon tanks are unceremoniously dumped, waiting at the curb to be buried at a local landfill. Is there any alternative?

If a water heater cannot be repaired—although many times we've repaired ones with small leaks with metal epoxy designed for use on engines—then here are a few practical uses for old tank.

As mentioned in chapter 5, we have used the thin, sheet-metal outer shell of a water heater tank for one of our various compost containers. This worked adequately, though there are better and more convenient compost bins. The water heater shell rusted readily in the yard, and the edges were sharp; it was also slightly too tall for easy use, although when we dropped scraps in the top we could gradually remove compost from the bottom. But due to its less than attractive appearance and unwieldy height, we used it for about a year, then emptied it and recycled the metal.

A cleaned-out inner tank with the shell and fiberglass insulation removed can be set in a corner of your garage and filled with water. (Clean the tank by opening the drain hole and running a house through the inlet.) This is perhaps the easiest, most convenient, and cheapest way to store large amounts of water in the event of an earthquake. Often, you can still use the spigot (the clean-out valve) designed to drain water from the tank.

If properly plumbed and safely situated, you can use an old tank as a wood-burning water heater, a possibility already discussed in the energy chapter, where we also discuss the potential for using old water heater tanks to construct a basic solar water heater. If you are using such heaters only temporarily, you can simply fill the tanks with water, let the water heat up from the fire or sun, then drain the tank for wash or bathwater. For more permanent installations, see chapter 7's discussion of plumbing options.

## PLASTICS

When plastics made their debut in consumer products in the 1940s, they were considered an imitation of "real" materials such as wood, paper, and metal. Plastic used to be synonymous with ersatz, the epitome of cheap and disposable. Today, plastic technology has advanced so far that some plastics are far superior to other products: for instance, Lexan is bulletproof; linear-D grocery bags can hold a surprising amount of weight; and some plastics actually burn more completely and cleanly than wood. Certain plastics are now lighter and tougher than a comparable amount of wood or steel. Petroleum-based plastics have come into their own, and—barring unforeseeable events—they seem to be here to stay.

Only in the late 1980s had serious efforts begun to recycle plastic products. Yet in most communities, only the one-quart (or liter) plastic soda pop containers are routinely recycled at local recycling centers. Gradually this is changing. Some supermarkets and dry cleaners will now accept returned plastic bags, and a few manufacturers have taken "junk plastic" and fabricated such items as flowerpots, parking lot stops, toys, shoes and other clothing, as well as posts and decking that looks something like wood. These are all steps in the right direction.

What are some of the ways we can make use at home of plastic products that would otherwise be dumped in a landfill or, worse, incinerated? As with recycling metal, paper, or other materials, the only limit is really your imagination.

SCARECROWS. Used plastic trash bags make excellent scarecrows. In your garden, set an approximately six-foot post, about the thickness of a broomstick. Place a plastic bag over the pole, and tie it off toward the top as well as once again a little lower on the post, at about waist level on the "torso." If you tie off the plastic at the right places, this shape will actually appear to be someone standing out there. Interestingly, this is much more effective in keeping unwanted critters out of the yard than the traditional scarecrow. Even with a light breeze, the flaps of plastic flutter in the wind and make noise. The noise and the constant movement of the loose parts of the plastic tend to frighten birds that might otherwise linger in the corn patch or garden. (This idea was sent to us by Edson Johnson of Sunland.)

SANDWICH BAGS. Heavy-duty bags should be washed, hung out to dry, and reused. There's no reason to use these just once and then toss them. There are attractive drying racks available from catalogs, or you can make a dryer with a short piece of wood whose branch stubs have been left on.

Some of the lighter-grade plastic bags have a shorter life span. One way to press these into service is to wear them as "gloves" while painting; assuming that the bag is large enough, simply place it loosely over the hand and secure the opening around your wrist with tape, a rubber band (not too tight), or piece of string. You'd be surprised how much freedom of movement you still have with your hand in a bag. When you're done painting, simply remove the bag and your hands will be clean.

These recycled plastic gloves also come in handy when cleaning guns or working with abrasives; be cautious (test first) when using plastic gloves with solvents, because some solvents melt plastic (and that could result in a burn).

MESH BAGS. Often grapefruits, oranges, or onions, and sometimes potatoes, come in netlike bags made of plastic mesh. These bags are sturdy and allow an airflow through the contents. They are useful as laundry bags.

A friend from Altadena carries a few of these bags when he's camping. After meals, he washes his pots and pans and dishes, and then, rather than laying his gear over rocks, he puts it all into mesh bags. He rinses the entire contents, and then hangs the bag from a tree to drip-dry.

Here's another use for those mesh bags. If you know of a steadily flowing stream and a place with plentiful oak trees, you can peel acorns, put the acorns in the bag, and put the bag in the river. Over the course of a week, the water will gradually wash out the tannic acid so you can eat the acorns or grind them into flour.

BLOWN FOAM. Plastic foam cups, plates, egg cartons, meat trays, and packing "peanuts" are now found everywhere. Originally they were regarded as especially modern and convenient, but now they're just "normal" rubbish. By the way, although this material is widely called "Styrofoam," that name is a registered trademark that applies only to those products made by Dow Industries.

Rectangular plastic foam trays can be washed and reused. They are often suitable for serving toast and other hot side dishes at our dinner table. They can also be used to store things in the refrigerator, or depending on their size they can be placed under flowerpots as a water catch.

And these rectangular trays can be cut into triangular sections and used as markers in the garden, because they're completely waterproof, although they will eventually degrade in sunlight. You can get three or four triangular "flags" from one tray. Press the pointed end into the ground and on the wider top end of the marker, write the name of the seeds just planted.

Plastic foam egg cartons make good protectors for young trees. Timothy Hall, our former neighbor, has tied as many as ten of these around a tree trunk. He did this to prevent gophers from chewing on the soft bark. He has also used egg cartons to protect a tree from sunburn when it had to be transplanted out of season. Taking a small pocket knife, Hall cuts out some of the inside dividers until he can close the carton, vertically, around the young tree trunk. Then he secures them in place with a bit of wire or string.

We've heard of people using plastic foam cups and egg cartons in their walls or ceilings as a sound insulator. This works well and is best done as part of a construction job so that all egg cartons and cups can be stapled or strapped between the studs inside the walls, where they are neither visible nor taking up your valuable space. One time Christopher installed plastic foam egg cartons while doing a wall renovation where bees had been removed.

PLASTIC JUGS. Many products are packaged in 1-gallon plastic jugs, including milk. The jugs that contain liquid bleach and laundry detergent are the most durable, very useful for water recycling (as discussed in chapter 6); you can save your "warm-up" bath and shower water in these jugs, then use that to water your houseplants or garden plantings. Or instead of sending it all down the drain, using these handy jugs you can save some of the graywater left after bathing or dishwashing for irrigating plants. Also, by pouring two gallons simultaneously into a toilet bowl, you can flush the toilet using water you've already paid for.

One of these plastic jugs can be used to store your toilet bowl scrubber. Simply cut out a circle on the upper side of the jug, leaving the handle, and keep the jug and brush behind the toilet. By creatively modifying plastic jugs with a utility knife, you can also fashion funnels and scoops. For instance, to make a funnel, cut the upright jug horizontally around its "equator," then use the top half inverted, with the spout for the funnel's drain. To make a scoop, orient the jug sideways and cut off the upper half, leaving the handle along with a scoop-shaped lower section of the jug. To avoid unnecessary mistakes, first draw the cutting line with a crayon or marker. With a clever job of cutting, you can make one jug serve as both a scoop and a funnel.

Plastic jugs make good floats, too, in swimming pools, lakes, near boat docks, and even along the ocean shore, to a lesser extent. If you need to mark off a swimming area from the deeper water, use several plastic jugs with screw-on tops to trap air, then tie them together with a long cord.

FOOD STORAGE TUBS. Plastic margarine and yogurt tubs with their tight-fitting lids are excellent storage containers. After we extract honey from our beehives, we pour the melted beeswax into these tubs. They are also good for leftovers, for storing grains and seeds, and for carrying pet food while on trips.

BREAD BAG CLOSURES. One time while camping we noticed that some other hikers who had washed clothes had hung the laundry out on a line, securing each item with one of those little plastic squares used as bag closures for store-bought bread and produce. We were surprised: who would have thought that something so seemingly useless could provide such a service? These are ideal for campers because they are light and take up hardly any space. We've used them at home in a pinch, though if left out on the clothesline in the sunlight for any length of time they become brittle and easily crack, probably because they are made of polypropylene, a form of plastic that is biodegradable.

BERRY BASKETS. Birds will often pick at newly transplanted or sprouted seedlings. Undoubtedly, these are a delicacy. How does a gardener protect plants from hungry wildlife? Plastic berry baskets often do the job quite well. Invert one basket over each sprout and use a small pebble to hold the basket in place. For added strength, we have used two berry baskets together.

OLD GLOVES. When plastic or rubber gloves are worn out and ready to be discarded, you might still be able to make use of the fingers if you cut these off and stretch them over the ends of wooden-handled tools such as trowels, brooms, rakes, shovels, and so on. Over time, these wooden handles tend to dry out and begin to crack and splinter. Those plastic glove fingers help protect your hands as well as providing a good grip and protecting old handles from further deterioration.

FILM CANISTERS. Plastic 35-millimeter film containers are excellent lightweight and watertight containers that have dozens of uses. Most photographers just toss them away, and almost any photo shop would be likely give you all you want, for the asking.

These little containers measure only 2 inches in length by 1¼ inches in diameter. The snap-on plastic lids assure a waterproof seal. We hear of new uses all the time and will include here just a few that we've personally tried.

Campers and hikers are especially concerned with efficiency in size and weight, and film containers are ideal for storing dried spices and condiments such as salt, pepper, coffee, tea, oil, vinegar, and any of the other ingredients that help to spice up a backwoods meal. These provisions can also be used at home, kept in a drawer or cabinet. Be sure to label the visible side of the containers, depending on where they will be stored, so you don't need to open every one to find what you're looking for. Small nonfood items can also be packed in these containers, for example, safety pins, fishing hooks, sinkers, loose change, spare keys, a sewing kit, and so forth.

These containers also make good carriers for wooden matches, but you will probably need to cut off the bottom of each matchstick so that they will fit. And to ensure that you'll be able to ignite the matches, cut a small round circle of fine sandpaper and paste it into the lid. Obviously the matches should be placed in the container "heads down" to avoid rubbing against the sandpaper.

For the traveler who carries a toothbrush, a film container is the perfect way to keep that brush clean. Cut a small rectangular hole in the lid, push the toothbrush handle through the lid from the inside, then snap the lid onto the container.

For bow hunters or archers, a film container is a safe place to store razor-sharp broadheads (hunting points) until they are needed. The broadhead goes into the film container point first, and a small hole in the cap will secure the threaded end in place.

## El Viajero

Christopher was once challenged to devise a miniature survival kit, the contents of which had to fit into one 35-millemeter film container After exploring possible contents for such a kit, he finally came up with what seemed to be an ideal combination.

We called this tiny kit El Viajero, Spanish for "traveler," because it could travel anywhere. Here are the provisions that we have selected for inclusion:

- *one natural bouillon cube (for broth, or for flavoring wild plants);*
- *eight small wooden waterproof matches, which were*
- *wrapped in 3 yards of 10-pound-test fishing line;*
- *four paper clips (the wire would have various uses);*
- *two small safety pins (which have many uses);*
- *four rubber bands;*
- *four small fishhooks;*
- *a small wad of cotton for starting fires (real cotton, not the synthetic "cotton" that doesn't burn readily);*
- *a round piece of sandpaper stuck into the inside of the container's cap (for striking matches);*
- *an approximately eight-inch square of aluminum foil, folded down and compressed, which could be used as an emergency signal mirror or even a cup;*
- *and one single-edge razor blade (with the sharp edge taped to prevent injuries). That blade would serve as an emergency knife.*

We originally included a dime for phone calls, as well, but this is no longer necessary because you can now reach an operator or make emergency calls on any push-button phone without needing coins.

Amazingly, all this fit into one film container. We admit that this kit was somewhat of a gimmick. The mental and physical preparation of survival training are really *far* more crucial than gadgets. Still, this was an enjoyable and useful exercise. When assembling a survival kit, remember that a top-quality knife and a fire-starting device—we recommend the Doan's Magnesium Fire Starter—are most important of all.


PAPER

Trees give their lives for us! Our society uses vast quantities of paper, much of it made from virgin sources, and we want to add our voices to the chorus urging Americans to recycle paper: one ton of paper (about two pickup truck loads) recycled saves seventeen trees!

STATIONERY. We inspect all the incoming mail for usable note paper. Paper that is entirely (or mostly) blank on one side is saved in a special file. Our cup truly runneth over, a testament to the incredible volume of perfectly *usable* paper coming into every household.

We used to save narrow strips of blank paper and place them by the telephone for messages. However, we now get so many calls that we've found it necessary to keep a notebook by the phone to record calls. On occasion, we have stapled recycled papers together to create our notebook.

MIXED PAPER. Some recycling centers take "junk paper" or mixed paper, including glossy inserts, "boxboard" cartons, telephone books, and magazines. Whether or not they accept these types often depends on the prevailing market price for secondhand paper. And with so many cities recycling paper now, the price of newsprint, for example, has dropped steadily. As a result, "junk paper" commands steadily less income, and some recycling centers don't consider it worthwhile to process. Because in many localities citizens must pay for all trash that is collected, it's probably worth your while to look for each and every way possible to reduce the household waste stream by turning rubbish into recyclables. Call your local officials and recyclers to determine who is currently taking what kinds of paper. Because our city now has a curbside recycling program, any paper products that we don't reuse can be given to the city for recycling.

"RETURN TO SENDER!" Some advertisers send their missives continually, whether or not you want anything to do with what they are pitching. We've heard other people state that they enjoy burning "junk mail," and even that they solicit more to burn in their fireplaces. We consider that an immoral and unnecessary waste of paper. A better choice is to return the letter and request that you be removed from their mailing list.

We return unwanted mail in two ways. Either we write "Refused— Return to Sender—Remove from Mailing List" prominently on the outside of the envelope, then drop it back in the mailbox, or we use the senders' postage-paid return envelopes to mail all the contents back to them, always being certain to write "Remove from Mailing List" next to our name. Because companies and organizations pay a higher rate to receive those postage-paid envelopes, this extra expense probably makes an impression.

Our daily mail processing rarely takes more than ten minutes. Making an effort to separate reusables from recyclables, and also to stem the flow of advertising mail, has enabled us to become a part of the solution, not a part of the problem.

BURNABLES. Many types of paper have coatings or bleaches that can release dioxins and other poisons when burned, but we use paper that we know to be safe to burn—newspaper, uncoated paper, cardboard—as kindling in our fireplace. Be selective about burning paper; many communities have laws against burning trash.

Here are some of the other ways in which we work to make the best use of all the paper-related resources that we handle daily.

ENVELOPES. Our first exposure to envelope recycling came when Christopher was getting acquainted with the folks at WTI. Geraldine Hogeboom showed everyone at one of the meetings how to reuse an envelope. First, using a pencil or letter opener, she slit open the seams of an envelope. Then she refolded the envelope, inside out. Some envelopes have enough leftover glue so that you can just lick the seams and seal them. Others require two or three pieces of tape (or swipes with a glue stick) to be secure. Such an envelope looks unusual at first, but it is clean and certainly usable.

Many advertisers send return envelopes for use when you buy their product or pay a bill. Most of these free envelopes can be recycled and reused creatively, but some take more time to prepare than others. We currently save only those envelopes that are easily reused with either a blank stick-on label or with a few strokes of a felt-tip pen. We affix stickers to these envelopes that say on each envelope in red ink: "Save Our Trees!! Recycle ALL Envelopes!" This encourages the recipient to recycle as well (and will also deflect any errant thoughts that we're just cheap or sloppy).

Admittedly, in the many years that we've done this, we've personally "saved" perhaps only one-half of one tree. Nevertheless, if each of us made a practice of saving this much of the paper that passes through our hands in our daily lives, the savings would be considerable.

Those envelopes that require too much work to reuse we put into a bag with other paper to be recycled by the city.

FROZEN-JUICE CONTAINERS. Those cylindrical containers for frozen fruit juice are handy for candle makers. Since they are usually made of thick cardboard with a nonstick coating inside and a metal base, they are ideal for small candle molds.

Begin by pouring a little wax into the container. Next, attach two strings or lightweight sticks over the opening of the juice container, arranged like an X; gently secure your wick to the middle of this X so that the wick hangs down in the middle. To make the best candles, it's better to pour in about one inch of wax at a time. Wait at least an hour between pourings to prevent air pockets. When finished filling the container, wait at least twenty-four hours for the wax to harden, then peel off the cardboard. Your candle is finished!

WAXED CARDBOARD MILK CARTONS. The quart- or pint-size waxed cardboard containers that contain dairy products or kefir make good temporary planters, and these can be planted directly into the ground when the plant is large enough to be transplanted. The bottoms should be punctured to permit roots to grow out.

Here's another excellent idea, especially for those of you who frequently buy products in this kind of container. When empty, squash the container flat, then fold the bottom over once or twice so that you have a squat square of cardboard. Store these folded-down squares in another of the same type of empty container, but not flattened. You can get about twelve of the flattened cartons into one empty quart container. Since these are waxy, they are excellent fire starters. You can use them in your fireplace, wood stove, or outside barbecue. During wet-weather camping, you can also carry one in your pack to help get an outdoor fire going.

PAPER TOWEL AND TOILET PAPER ROLLS. Both of these types of cardboard roll are useful as candle molds, using the same technique described above. Nathaniel Schleimer of Pasadena once exhorted us *never* to throw away these useful rolls. He had experimented with at least seventeen uses for them, not all of which turned out to be viable.

For instance, these rolls are practical for storing the excess sections of long electrical cords. Slip the roll over a rolled-up section of the cord to keep it from getting underfoot.

Or the rolls can be used as napkin rings. Cut them into smaller segments and then cover them with aluminum foil, wrapping paper, paint, or just use them as they are.

On his desk, Nathaniel had a unique pen and pencil holder, which he made by mounting four rolls of different heights vertically on a wooden mount, then spraying the whole thing with paint and sealing it with shellac.

Nathaniel also showed us that the paper tubes work well as mailing tubes for items you don't want folded; as small percussion instruments (seal the ends after putting noisy little objects such as beans, lentils, BBs, or pebbles inside); as a way to keep your ties or socks sorted; and as a shelf organizer for keeping small hand tools together. When we were visiting Nathaniel in his "bachelor home," he said to us, "How easily could you manufacture a hollow tube like this? Just think! What a *very* useful material, and basically free. You'll find many more uses for them if you just keep your mind open."

ICE-CREAM CARTONS. A half-gallon cardboard ice-cream container makes a great biodegradable planter. You can pot an avocado in it, or a sprouted potato, and then just plant the whole container when the seedling gets big enough.

BROWN PAPER BAGS. Are these paper grocery bags an endangered species? Most supermarkets baggers still say, "Would you like paper or plastic?" but some now say "Is plastic okay?" A few go right ahead and bag your goods in plastic, providing the more expensive paper bags only on request. And some stores actually request that you bring clean bags to them so that they can reuse them.

Who doesn't have at least a dozen uses for a brown paper bag?

We've used brown paper bags as make-shift vacuum cleaner bags, and they've worked fine, all manufacturers' warnings to the contrary. On one particular vacuum cleaner we had, the largest size of brown paper bag was too big, so we used one size smaller. Experiment with sizes, depending on the construction of your machine. To secure the bag in place, we fold back the top of the bag so that there is a snug fit, then wrap it with thick masking tape so it won't come off. A rubber band could also work.

Few people ever consider the ramifications of millions of householders throwing away the dirt and dust from in and around their homes. To some extent, household dirt and dust is already soil or potential soil replenishment. To throw it away is therefore a depleting process that ultimately is an incipient stage of desertification. The contents of a full vacuum cleaner bag are perfectly acceptable to add to a compost pit, worm farm, or as a top layer of mulch for potted plants. And if you want to discard the entire bag with its load of dust and lint, dig a hole in the compost pit and bury the whole bag. It will be fully decomposed in a matter of weeks.

At an outdoor event in summer, we once saw a very lovely use of brown paper bags. There was no moonlight, nor were there any electric lights along a particular outdoor trail, so the organizers of the event lined the pathway with "luminarias," which consist of an open brown paper bag into which is placed a candle in some sort of protective container. For example, some of the candles were in votive glasses, and others were placed in cans—soup-size or cat-food-size cans. A few pebbles or a cup or two of sand placed into the bottom of each bag will prevent accidental fires if a bag were to be blown or kicked over.

Beautiful patterns and figures can be cut into the sides of the bags, as well. If the event is a Halloween event, you can cut a face of a jack-o'-lantern or witch into the bag. If it is a birthday commemoration, you can cut a letter into each bag to spell the celebrant's name or some other greeting. These glowing lanterns are quite attractive. Obviously, you wouldn't want to use them in dry conditions during a windy night, and you'd want to keep them where they can be easily monitored for safety. Just to be safe, set the bags on fireproof surfaces such as bricks, gravel, concrete, or asphalt driveways.

NEWSPAPERS. Whereas everyone has at least a dozen uses for brown paper bags, everyone has at least a hundred good uses for newspapers. Where do we start? This list is of course only a beginning.

Newspapers make an excellent undermulch (sometimes called "sheet mulch") in gardens; they will stop the invasion of grass and provide a base for productive soil. Lay out the papers over those areas you want to mulch, then cover them with materials that will hold them down: wood chips, grass clippings, leaves, pine needles, compost. Newspapers decompose rather quickly if wet.

There has been some controversy over whether or not the ink in newspapers can be harmful to your garden. Much of the concern has focused on the colored advertising and comic sections. One response to these questions has been to compare the amount of ink in the paper with the actual weight of the paper, in which case the ink turns out to be pretty insignificant. Then when you consider the amount of potentially harmful substances in the black or colored ink, the proportion is quite minuscule. We've read several reports on this and have concluded that mulching with newspapers will not result in harm to your soil. Fortunately, many magazines and newspapers are apparently now printed with vegetable inks, and it's an easy matter to call up your local paper and find out.

The comic sections of newspapers make good place mats at children's parties. Often children won't even notice an "adult" place mat, but clearly enjoy the colorful comic sections. When the children are done with their meals, you'll have no place mats to wash. Just fold these up and put them in your newspaper stack for the recycler. We know people who use the brightly colored comics and advertising sections for wrapping paper, as well.

Newspapers also have a good reputation for window washing, since they clean glass well and don't leave any lint. Many of us will continue using them rather than going out and buying special cloths for this job.

During the late 1960s and early 1970s, several companies sold "logrollers." These were devices that helped you roll up some newspapers into a "log," which could then be burned in the fireplace as easily as real logs. Unfortunately, they didn't burn very well at all. The better versions were designed so that there would be a hollow channel in the middle, and these burned slightly better than a solid roll. But one could just as easily roll some newspapers around a broom handle, tie them up, and then slide the paper roll off. The point is, these rolls only burn well when added to an already well-established fire.

A far better idea for burning newspapers is a log press. With this device (we've seen three or four models advertised over the years), you first soak your newspaper in a bucket of water—ideally, for about twenty-four hours, though we've had good results with about thirty minutes of soaking. Once soaked, you have a gooey, pulpy mass. You fill the rectangular box of the press with this pulp, then press down (in some models you press down by hand; others are designed for foot pressing). Once you've squeezed out all of the water, you release the wet "brick" and lay it out to dry. Drying time depends on air temperature and time of year but generally should take no longer than a week.

These newspaper pulp bricks burn quite well! We'd highly recommend them, despite the fact that *Consumer Reports* magazine rated them poorly. Part of the reason for that low rating is that the testers used a misleading premise: they measured the BTU (British thermal unit) output of the newspaper bricks and suggested that the work required to produce this amount of BTUs wasn't worth the effort when compared with the BTU content of commonly used types of firewood. But comparing newspaper logs with firewood wholly misses the point! The better question would be: does this device render newspaper into an easily combustible product, and is operation of the device practical? Our answer to both is a resounding *yes*, especially considering that the raw materials are free and readily available.

Moreover, this is an excellent way for youngsters to earn money. They can collect newspapers, make burnable pulp bricks, then sell these to neighbors with fireplaces. When Christopher worked as a counselor at Summer Kids day camp in Altadena, he would occasionally bring the log press for campers to use. The children loved to soak the newspapers in water, fill the press, and compress the pulp to force out the water. And each child always wanted to take home a sample of the results. With television vying for our children's minds with ever greater sophistication, it is a marvel when any such old-fashioned gadget can hold their attention.

Even if you can't think of any other way to reuse your newspapers, *do not* just toss them into a trashcan. More and more cities and towns are implementing recycling programs, and newsprint is one of the easiest materials to recycle. Put out the newspapers on recycling day. If you don't have a curbside recycling program in your city, give the newspapers to neighborhood children for school fund-raisers, or take the papers to a local recycling center.

Telephone books. Although in some places they are not accepted by newspaper recyclers, old telephone books can be torn apart and the pages used as kindling for fires or as mulch. As mentioned in chapter 5, Ernest and Geraldine Hogeboom of Delhi, California, have made mulch by ripping the pages out in increments of about fifty pages, and laying these down in overlapping stacks in those areas of the garden they wished to mulch. They reported that the pages of the telephone book are generally decomposed within a month.

An old telephone book can also be an ideal hiding place for money or jewelry. All you need to do is cut open a cavity inside. We've even seen some of these "hiding books" with pins or glue to keep the cut sections of the book together.

Here's another idea: when we need a backstop for target practice, we use telephone books. Having put about five old phone books into a square 5-gallon can, we fill the can with water and the telephone books swell up. Assuming that it's safe (and legal) to be shooting in your neighborhood, these are ideal targets for use with an air rifle in the back yard or for use with a .22-caliber rifle out in the desert or in some other safe location.

A similar kind of backstop can be made for archery. We've seen bundles of telephone books tied with wire and stacked on top of 55-gallon drums. One acquaintance of ours would practice with his bow by shooting down the length of the driveway. Using both field points and hunting points, he shot into the sides of stacks consisting of approximately twenty phone books. Although he rarely missed the target, those few arrows that did miss would be stopped by a wide sheet of plywood secured to the outside of his garage door.

CARDBOARD. There seems to be no shortage of cardboard, especially all the boxes that are continually discarded by stores and supermarkets. David Ashley of Ashley Enterprises, working in both Los Angeles and Santa Barbara, makes his income by recycling cardboard discarded by stores.

In addition to the countless ways of using cardboard for storage, large sheets of cardboard are particularly effective for mulching sections of your garden where there is persistent, hard-to-eradicate grass. It's best to cut the grass as short as possible, cover it with several layers of cardboard, and then cover the cardboard with a few layers of organic materials or compost. While this may not totally eliminate undesirable grass, the mulch layer will enable you to grow garden plants and over time will assist in improving the health of the soil.

BITS AND PIECES. Odd bits and pieces of paper or cardboard can be saved in a paper bag and used as a "starter" in a wood stove or fireplace. Be sure to take wood stove ashes and sprinkle them around your fruit trees and garden, because they are a valuable source of nutrients for your plants.

## CLOTH AND FABRICS

Clothes that can still be worn by someone ought not to be tossed into the trash. Usable garments that you no longer wear or that you've outgrown can be taken to one of the many organizations that provide clothing to homeless people or to less fortunate people here or in other countries.

Expand your thinking from "I don't want this anymore" to "I'll bet someone else would *love* to have this." The shirt or warm coat that you donate may make a world of difference to a homeless person. Of course, once you get used to recycling clothes, you need to be honest with yourself when tempted to donate something that is truly worn out and unlikely to be appealing to anyone.

It is always better to use a garment to the full extent of its useful life than to toss it away due to the vagaries of fashion. Some people will really need to stretch their brains over this one, because today clothes are often purchased as body adornments, but historically speaking, the marvelous techniques and technologies that brought us woven fabrics have been in use only a relatively short time. Not many would want to go back to animal skins. So, as with so many of our seemingly abundant modern resources, fabrics and clothing are valuable resources that should be used wisely. Needless to say, proper care for one's clothes will extend their useful life. For example, Dolores has long lamented the fact that moths eat holes in her good wool clothes. By careful comparison shopping, she recently purchased a secondhand cedar chest, in good condition, for about a third of its value. The cedar chest will pay for itself many times over by saving wool clothing.

We ought to make and/or purchase clothing granting the highest regard to quality, longevity, and practicality. Yes, beauty is important, but there's no reason why beauty must be achieved at the expense of genuine quality.

Besides donating used garments to a thrift store or charity, what other uses are there for the cloth and fabric items that we periodically discard? Remember, in addition to clothing we have bedding, towels, carpets, draperies, and various other items that have any number of uses.

BED LINENS. Cotton sheets that are no longer usable for bedding can be pressed into other types of service.

Cut into approximately seventeen-inch squares and hemmed, old sheets make great kerchiefs, which have many uses for either the camper or city dweller.

Clean cotton sheets can be cut into variously sized strips and kept for bandages. You can cut squares, short strips, and long strips, fold them neatly, and add them to your first aid kit.

A round piece of cotton can be cut and sewn into a cone shape to make a reusable coffee filter. We've used one of these for several years, and it's much better than discarding a paper filter each and every time you make coffee.

Cotton sheets, bedspreads, or other durable fabrics can be rolled into a long "snake" (sewn along the length, if filled with sand or beans, for weight) and used as weather stripping at the base of doors. There are store-bought products that do this job of blocking drafts, but good substitutes can be easily made from scraps.

Those who know how to weave can take old sheets and bedspreads and make them into serviceable hot pads, washcloths, "rag rugs," and dog beds. This style of weaving on small looms is popular in traditional cultures, such as in Mexico. If you juxtapose a variety of colorful fabrics, you'll get very attractive results.

CARPETS. Insoles for shoes and boots help to keep feet warm and dry, and insoles provide a snugger fit for shoes that are slightly too big. An excellent insole can be made by cutting a piece from an old carpet. Low-pile (not shag) carpet works best. You can trace your feet with a marker, then cut the shape to fit using heavy-duty scissors or a utility knife.

Sections of old carpeting work amazingly well as a mulch, even in areas where you are battling wild grasses and persistent weeds. We recommend first scything down the existing growth, though this isn't absolutely necessary. Lay the carpeting down, overlapping at the edges, and leave it there. In some areas, perhaps for aesthetic reasons, you might choose to leave the carpets in place for only a few months. By that time you will probably have smothered most of the plants that were growing underneath, and can very likely begin gardening or landscaping in that area.

On hillsides, carpeting will help to prevent erosion after rains. Some plants will eventually start growing through the carpet and hold the slope in place.

Strips of carpets laid along the paths of outdoor walkways help to hold in moisture, preventing slippery spots in the wet season and in some cases actually contributing to improvement of the soil below. Of course this soil improvement benefit depends on what type of carpet is used. Wool carpets or runners made of organic materials such as jute or sisal are much better for the soil than petroleum-based synthetics, though the latter still have some value for mulch and erosion control.

*Outdoor Dojo Mats.* During our time of getting acquainted with WTI in the late 1970s, we were surprised to see the workout area where the survival training classes were conducted. This was a flat area approximately thirty by thirty feet, which was entirely covered with old carpets, large pieces discarded by carpet layers when they'd installed new carpets. As each layer of carpet became old and began to deteriorate, new layers were simply laid on top. In time, regardless of the original color, all of the carpets faded to the same gray or beige color.

Although the workout area was outdoors, no coverings or protection were ever provided for this carpeting. All that was required was for someone to sweep once or twice a week, particularly if the weather had been windy. After rain, the carpeting tended to dry within a few days.

This was an excellent padding upon which to conduct the classes, the equal of those tatami mats used in most martial arts dojos. Even when they were wet, working on the carpets was far better than working in the mud. Some of us would lay down sheets of plastic if we were practicing postures or movements that required lying on the ground.

When we moved into our home, the existing carpeting was old and damaged. We removed the carpets and laid them over a large, flat section of the back yard to make our own exercise area. Now there are probably five or six layers of carpet there.

OLD SHOES. When Michael Rubalcava (Ellen Hall's son) was ten years old, he had a small nursery of "survival plants" at the WTI headquarters in Los Angeles. He grew aloe vera, herbs, prickly pear cactus, and many other useful plants. He used recycled cans, plastic containers, and ceramic pots for potting his floral beings. He also used old shoes.

He saved his own shoes and collected others, filled them with soil, and planted seedlings in them. We often laughed about these, yet they always sold well. The novelty factor was apparently very appealing. In fact, most shoes make stable, long-lasting "pots." They're attractive, and they're good conversation pieces.

When we sold plants at the local farmers' markets, we would take at least one plant potted in a shoe as the attention getter and as a testimony to clever recycling. Many people would stop, point at it, and laugh, and nearly always, someone would buy it.

OLD BLANKETS AND DRAPES. There are many uses for old blankets and drapes, which tend to be made of durable fabric, including seat covers for cars, mulches, and sunshades. The rag bag should be the last resort, and the trashcan should be out of the question.

On one of our early wild food outings in the San Gabriel Mountains, we met Christine Zellich, who wore a pack that she had made from old drapes. We were first attracted to the pack because of its bright colors, illustrated with floral patterns of Hawaiian origin. But as we examined it, we saw that it was a homemade pack.

Christine said that she first made a simple pattern out of paper, and folded the paper to make sure the pieces would fit together as she intended. Then she cut sections from the old drapes. The body of the pack was one folded piece of the drape, with two rectangular side panels. She also made a closure and carrying straps. It was really quite simple. Due to the strength of the original curtains, this was a sturdy day pack.

## CHRISTOPHER'S TALE OF DUMPSTER DINING

When I was young, my mother would tell us, "Eat all the food on your plate. People in Asia are starving." Although I supposed that cleaning my plate would in no way affect Asians' starving stomachs, I finished every morsel of my meal. I knew that what my mother meant was that we should be grateful to have food when others in the world were hungry.

Since then, I have seen with crystal clarity that we are not only a nation of plenty, but a nation of wasters. I had a most upsetting experience a few years ago after attending an outing up in the San Gabriel Mountains. A fellow teacher, Drew Devereux, had shown his students how to recognize edible wild plants that had sustained generations of Native Americans. After we dropped off the students, we were standing near the LaSalle High School's large trash bin talking. We looked in the bin, and what we saw shocked us. The trashcan contained still-wrapped sandwiches of various types, lots of fruit, and even some unopened potato chips. Oh, sure, we had heard of Charlie Manson and his clan scavenging from trashcans to survive, but this was different. Here we were being freely offered an unexpected gift, which we joyously accepted. Yet we wondered—how can this be? How can all this perfectly good food be just thrown away?

Since that eye-opening day in my naive past, I've become acquainted with a fleeting but ever persistent population of trashcan food collectors, since I too have begun from time to time to check in the rear of supermarkets for the good food that is routinely discarded.

Among the younger generation, these folks who survive on the discards in supermarket trash bins are admired as near heroes who have beaten the system artfully. But to most of society, they are either pitied or scorned as bums, parasites, and scavengers.

Just who are these people whose hands reach daily for a slightly bruised tomato or for those potatoes that are too large to sell? Are they young, old, rich or poor, male or female, employed or unemployed? The answer is yes to all of the above, with an emphasis on elderly, widowed, fixed-income women. But I have met all types at the dumpsters.

For example, a late-model bronze Cadillac pulls up behind the supermarket. I have already collected two boxes of old tomatoes, celery, radishes, cucumbers, potatoes, and oranges that for one reason or another are deemed unfit to sell. Out of the Cadillac steps Saul, well dressed, middle-aged, smiling. We have never met before. As we both inspect the trashcan, we jokingly discuss what's on tonight's menu. He tells me that after his wife died, he traveled over most of this Land of Plenty. He comments matter-of-factly that he only occasionally has needed to purchase produce.

"Why should I?" he asks with a tinge of guilt in his voice, "when they throw away this kind of stuff?" He holds up a huge tomato and laughs.

But probably more typical of the trashcan survivalists is Paula, five feet tall, soft-spoken, in her late sixties or early seventies, a widow on Social Security, and with a face of a thousand wrinkles. She ambled around to the rear of the grocery store. Her timing was perfect, for she has learned when the produce department throws leftovers into the dumpster. I was there first and have already gathered much of the better discards on this rather scanty day. She tells me that she is gathering food for her chickens, but I can tell by the tone of her voice that this is not true. I tell her about the delicious meal I had the night before and ask her what she had for dinner the previous night. Realizing that there is no need to be embarrassed in my presence, she points into the trash bin, saying "Some of this, and I cooked some of these," holding up some carrot tops. I give her some onions and beets I've collected, and I dig around for a good head of lettuce. She carefully puts the produce into her wireframe pushcart and is looking for more produce as I turn to leave.

Because I regularly conduct wild food outings throughout Southern

California, I've seen the abundance of food available from nature. I can say without apology that I have had many meals made from this abundance and also from the abundance created by the wastefulness of store managers.

And wasted food in our society is not caused only by fussy grocers.

Dr. William L. Rathji of the University of Arizona surveyed trashcans in the Tucson area to see how much food is thrown away by individual families. From his Tucson data, he projected estimates of food waste nationwide. He figures that American families throw out 8 to 20 percent of the food they buy, at an estimated cost of $4.5 billion annually— almost as much as the federal government spends every year for food stamps and child nutrition programs. Rathji concludes that the average family wastes at least $150 worth of food per year.

"Homeowners go out of their way to save pennies at the store, but don't realize that waste of edible food at home cancels out that thrifty effort," says Rathji.

Collecting trash-food is a way of life for many in the City of the Angels. Maybe mothers should tell their children today, "Eat all the food on your plate, Johnny. People in Los Angeles are starving."

Both of us have continued to harvest thrown-away store produce. Our chickens love these greens, and so does Blue Girl, our goose. And with our help, Otis (our potbellied pig) even sent a thank-you note to Yosh, the produce manager of Pasadena's Hugh's Market. Yosh thoughtfully selected and boxed (from the discards) tasty fruits and vegetables for Otis.

*One of the scooters that Dolores rode to the store.*

*Christopher at work in the garage woodshop making twig pencils.*

*Dolores selling freshly gathered wild foods at a Pasadena farmer's market.*

*Otis eating pigweed (a.k.a. lamb's-quarters).*

*The rabbit hutch was built over a compost pit/worm farm.*

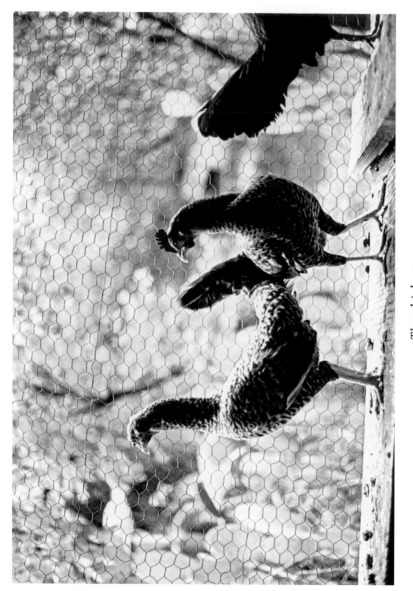

*The chickens.*

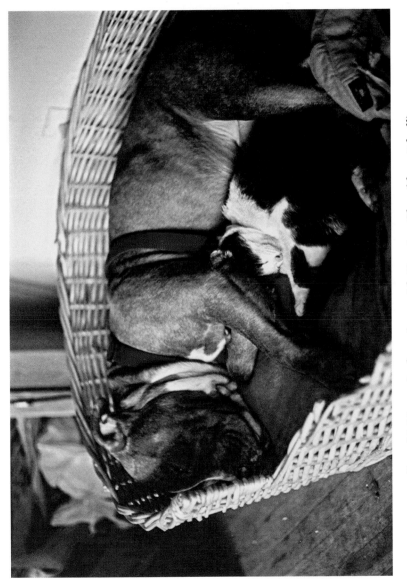

*Popoki (the cat) often slept with Cassius Clay (the pit bull).*

*A supply of canned food purchased at deep discount.*

*Christopher and Dolores harvest lamb's-quarter seed
(to be used in soup and bread batter) from front yard.*

*Dolores pinches off some mallow leaves for lunch.*

*Citrus trees grown for food, medicine, fragrance, and nectar for bees.*

*Dolores lays down cardboard mulch and
discarded marble pieces for a walkway.*

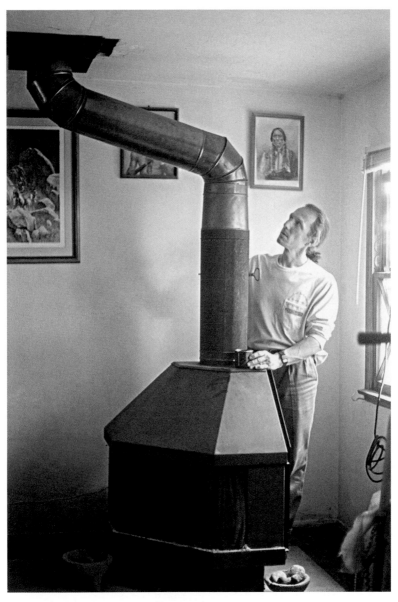

*Christopher inspects the newly installed wood-burning heater.*

*Ward Webb wires the panels to batteries.*

*Christopher's first "breadbox" solar water heater.*

*Mike Butler secures the solar water heating system.*

# 9 Economics and Self-Reliance

*How you spend your money really is a statement of what you believe in.*
—JENNIFER SCHUSTER, MINNEAPOLIS HEALTH WORKER, 1998

One cannot speak of genuine self-reliance without addressing the topics of money and economics. Financial considerations invade all aspects of our lives, and money itself is arguably an essential tool of survival.

Several generations ago, most people's way of life and means of support was farming. They farmed to feed their families, selling the rest for cash which they needed to buy machinery and other necessities—the various goods they couldn't barter or make.

These days, when it seems that *everyone* is constantly saying that they "never have enough" money, it's downright amazing how the rural folks of past generations could get by on the meager incomes they had. True, they lived on the land, and many of their needs could be obtained from the bounty of the land. Yet our ancestors' remarkable level of self-sufficiency was also due to the fact that many of them realized—as Thoreau wrote in *Walden*—that they were rich to the degree that they could do without certain material possessions.

We recently watched a television show in amazement, as a middle-aged couple with no children discussed their financial dilemmas. They described themselves as "poor," unable to keep up with their bills, worrying about whether or not they could make their house payments. Yet they were each earning well in excess of $40,000 a year. In the background, as they moped about poverty, we saw every conceivable electronic gadget and a house crammed with fancy furniture and other "stuff."

We are not without a need for money, but still we're rather surprised when we hear such stories, given how little money we get by with and how well we live. Let's consider how we earn and spend our income.

Most important, though we have both had "regular jobs" from time to time when circumstances necessitated, we have constantly strived to earn our income from our own talents, doing what we love to do, and doing what we feel is best, spiritually speaking.

In general, this means that our income is derived from teaching others what we do and what we know; publishing articles, booklets, and books about what we do and what we know; and selling crafts or products that we make and believe in.

Though neither of us currently has a teaching degree, we have found employment at nearly every college and recreation department in our area. Even at the local colleges, once we proved our competence in the subject matter we were proposing, we have taught semester-long courses on topics including integral gardening, wilderness skills, native American lore and religion, and writing for income. By contacting teachers and program directors, we have the opportunity to explain the areas in which we have some expertise, and as a result we are frequently called up to lecture in classrooms or to take students on field trips focusing on wild plants, Native American skills, or wilderness survival.

We both have a background in English, journalism, and creative writing, and so it has been relatively easy for us to take what we enjoy, what we believe in, and what we feel is important and share it with others via our writings. The challenge in being writers is that our writings have had to stand on their own merits. We have published several books and booklets on our own and with various publishers. Additionally, we have sold articles to *Mother Earth News, Back Home, Christian Science Monitor, American Survival Guide, Wilderness Way, Preparedness Journal,* the *Pasadena Star News,* and other periodicals.

Simply put, we have taken the time to discover what interests us— what compels us to get up in the morning—and further, we have made the effort to ascertain how we might support ourselves with livelihoods rooted in our own skills and talents. Although as a result of that choice we have had some very lean times, we have found this path infinitely more fruitful and fulfilling than getting an hourly wage in order to further somebody else's narrow ambitions.

For example, we have had our own natural gardening business. This was hard work but allowed us to put certain principles into practice in the real world. We built compost bins for most of the clients and fertilized their gardens with a mixture made from rotted seaweed and fish emulsion that we made in the back yard. But we would never just go and pull weeds or "mow and blow." Many people thought they liked the idea of "natural" yard care, but we did lose some customers when they saw what this meant, because they were actually more concerned that their yards look "nice."

We continue to organize courses in wild food identification, outdoor survival skills, and other topics of self-reliance. We conduct some of our classes in local wild areas. Some people in other parts of the country have the mistaken notion that Los Angeles is wall-to-wall cement and asphalt. Fortunately, we have not come to that, yet. Within easy reach here is the ocean to our south, the Santa Monica mountain range, Griffith Park (one of the largest urban parks in the United States), and the Angeles National Forest, which comprises the northern third of Los Angeles County. This forest area (not a forest in the normal way we think of that term, but rugged mountains) includes hundreds of miles of hiking trails at all elevations with all ecological zones represented, from riparian to desert. Just to the east is the San Bernardino National Forest, another nearby wilderness area where we sometimes conduct our classes.

At other times, we conduct classes in our own back yard, which also provides students with an opportunity to see some of the ways in which we practice urban homesteading. This teaching not only provides us with the opportunity to hone our own skills, but it has afforded us the unique opportunity to meet thousands of like-minded individuals over the past twenty-five years.

Ever on the lookout for ways to allow what we love to support us, we have registered our property and the adjacent property (which belongs to a nonprofit organization dedicated to "survival research and education") with the state's Department of Agriculture, which means that we are permitted to sell produce at certified farmers' markets. As described in previous chapters, we have collected the edible weeds from our land and sold them as "wild salad" at the certified farmers' markets in the surrounding communities ("certification" means that the state's agriculture department checks annually to confirm that we grow what we say we grow). We've learned that it is worthwhile for us to participate in the markets, which require a great deal of time and preparation, primarily in the spring, when wild foods are especially abundant. When we have a market stand we sell wild edibles, wild herbs (fresh, dried, and potted), crafts, cacti, and whatever else we have available.

As for marketing handicrafts, we began to explore this type of work when we formed a partnership with Helena Breese from Great Britain, whose company manufactured a unique "twig pencil," which they called "twigzils." These were made by inserting graphites into pencil-shaped tree prunings and whittling the writing ends to a point. Along with our several helpers, we made perhaps half a million of these during the three-year height of this business. When the interest died down, we looked for other things to make with the equipment we already owned— drills and sanders and wood burners and other tools.

We didn't look through the pages of those get-rich-quick magazines for ideas of things to make to earn a fast buck. Rather, we considered: What do we enjoy doing? What interests us? What would we feel good about making?

We began to manufacture bull roarers, ceremonial noisemakers based on Aboriginal designs; decorative gourds; hand drills and bow drills for primitive fire making; pump drills for drilling holes; elder-wood smoking pipes; wildcrafted herbal tobacco mixes; archery quivers and carrying cases from dried agave and yucca stalks; as well as baskets and many other kinds of crafts.

Though we did not always successfully sell the craft items we made, we learned important skills and discovered new ways to make what we ourselves needed, rather than being forced to buy yet more stuff from a store. In many cases, that feeling of independence was far more valuable than the few dollars we'd get from the sale of a particular handicraft. Plus, even when we did not manage to find steady markets for selling the crafts themselves, once we learned how to make certain objects, whether gourd bowls or capotes or tipis or herbal sachets, we were then able to conduct classes to teach others the skills involved in making these things for personal use.

GRETCHEN STERLING

*Dolores selling wild salad at Pasadena's Farmer's Market.*

Since that point when we really committed ourselves to being self-employed, we have worked steadily to make this path fruitful, and along the way we have developed both talents and a wide range of invaluable friends.

As a result, oftentimes income "comes to us" when someone calls and asks for advice on how to get some job done, knowing that we have many skills and contacts. For instance, someone might call to see if we can prune trees, perhaps remembering when we ran a gardening service. In the process, we'll end up with a load of wood chips and wood for craft projects. Or we get calls to do private consultations on any subject from gardening to herbalism and from wild foods to survival in extreme situations. We go to people's homes, and we provide private field trips. We charge for these services based on how far we need to travel and how long a consultation lasts.

We also get calls frequently to do talks and lectures on subjects such as wild foods and herbs, Native American skills, gardening, and metaphysical topics. Because we are nearly always encouraged to sell our books at such events, we can sometimes nearly double our income from the event when we add in the sales of books and booklets.

These are just a few of the ways in which we have discovered we can earn the income we need without slaving for someone else at an hourly wage. We began with a careful evaluation of our interests and skills, and then found ways to make these pay. Anyone who desires to do likewise should begin by making a comprehensive list of all your talents and skills, and also of all the products or services you could either barter or sell. Your list should also include those areas where you need to develop morally and spiritually. This gives you a very specific foundation to build upon, and from there you can start to make plans, then implement those plans.

As for marketing, we use every low-cost method we can think of to promote our activities, including sending out notices to the mailing list we've developed and maintaining the Web site created by Dolores. We put our newsletter, *Talking Leafs*, around town at health food stores and camping supply shops, and we explain to readers how to reach us in every article we write for other publications.

This way of earning an income is surely not for everyone, as it involves a fair amount of uncertainty and unpredictability, but we've found that it fits well with our beliefs and our ways of addressing the challenges of life. The "normal" way just doesn't seem to be for us. With a regular business, for example, we don't think that one can avoid certain compromises with one's values and independence. So many people feel compelled to do their work in a way that is not exactly right but merely "necessary" under the guise of various justifications. We do not wish to participate in such rationalizations if we have any choice in the matter. And we've discovered that we *do* have a choice.

## THE FOUR ILLUSIONS OF MONEY

During the early 1980s, we participated in WTI's monthly plenary sessions, which were conducted in Highland Park, California. These were all-day events where participants shared accounts of specific research they had been doing. Christopher had been giving presentations on a series of money-related topics, such as "What is money?" "What is the Fed?" "What is the IMF?" and so on.

The money-related lecture that stirred up the greatest emotional response was "The Four Illusions of Money." Christopher loosely based his presentation on an article by the same name that appeared in the winter 1979–80 *CoEvolution Quarterly*. The actual presentation, including discussion, lasted about two hours, covering many facets and dealing with comments and objections from the audience. What follows is a condensation of Christopher's key points.

When people are queried, almost everyone says that they do not have enough money, and would like to have more. Moreover, one of the most commonly cited goals given by people who work at jobs they dislike is to "make a lot of money." The reasons for this goal can be summed up in the following four rationales:

1. A lot of money will let me be free to do what I want.
2. People with a lot of money command more respect from others.
3. I need more money for my family.
4. Money is necessary for security in old age.

These are the four illusions of money—false perceptions of the world. These four illusions prevent us from seeing the nonmonetary truths about our existence and the choices that we make.

Let's explore them one by one.

*1. A lot of money will let me be free to do what I want.*

The way to see through this illusion is to make a specific list of all of your carefully considered goals. These can include travel, projects, achievements, possessions, skills, but not money—that is, not money itself as a goal. Next, examine the list and ascertain precisely how you can actually achieve these goals.

Yes, money can help accelerate the achievement of a goal. Still, once your goals are clearly established in your own mind—and clearly differentiated from "passing wants"—you can steadily go forward with steps toward the achievement of those goals. Money is incidental to this process and must not be allowed to determine the choices you make and the steps you take.

A large part of achieving any goal—perhaps the most important part—is to learn valuable life-enhancing skills via the active pursuit of that goal.

And many of the essential steps toward a goal involve working with other people. Thus, the more fruitful course than the pursuit of money is the development of strong trusting friendships, which means that you must be trustworthy and reliable yourself. This represents a path to freedom from the money god, and opens one up to truly life-enhancing experiences.

Remember, this perspective is proposed as an alternative to "going out to make enough money so I can be free to do what I want to do."

*2. People with a lot of money command more respect from others.*

This is demonstrably and abundantly false. There is no reason to assume that people with "a lot" of money automatically command genuine respect (they don't), or that people with "a lot" of money command respect because of the money. People who invite respect do so because of their personal qualities, talents, character, and experience. It may be that these qualities are the reason a person has been able to earn "a lot" of money, but money itself is not the basis for real respect.

*3. I need more money for my family.*

All too often people use this fallacy as an excuse for doing something they would rather not do. This rationale is especially typical of "breadwinners" who work extra hours and on weekends to buy possessions and vacations that they believe their families need and deserve. If you are getting more and more out of touch with your own family members because you are spending more and more time away from them supposedly in order to provide something extra, then you are falling for this illusion.

In any case, it would be far more valuable for everyone if these breadwinners instead spent valuable time with their family members, reorienting job and financial choices accordingly.

*4. Money is necessary for security in old age.*

Money is necessary in many ways, of course, but personal security, inner and outer, cannot be purchased.

The real security most needed by the elderly can be enhanced by money but can never be built solely upon money. Inner security arises with the development of deep friendships and with learning to be flexible and adaptable, for example, neither of which depend on money. In fact, one of the best ways to "prepare for old age" is to become the type of person, inwardly and outwardly, that others will want to be around and work with. This means being competent, helpful, flexible, honest, moral, curious, always willing to learn and share, generous, and so on— none of which are intrinsic virtues of the wealthy. Developing one's character is clearly one of the best ways to prepare for the calamities that might strike any of us at any age, even wars, depressions, and social chaos as well as a whole range of personal difficulties.

Obviously, money is an inextricable part of our modern life. As TV host Tony Brown has said, "If I've been accused of over-emphasizing money, it's because I place money right up there with oxygen as a necessity." Money is something that we cannot avoid and that we must deal with wisely and properly. And yet, we need to work hard, daily, not to be bewitched by the money god. If we live our life for money, we cannot but miss what life is all about.

Amazingly, there are many people with adequate money who feel "poor" because they mismanage their finances or because they hold on to unrealistic ideas about their world. On the other hand, there are many people who learn to separate personal illusions from actual needs, and in the process these people gladly persevere every day, in spite of a shortage (real or apparent) of material goods. Who is richer?

Perhaps the best first step in overcoming misperceptions about money is to "get small" in order to examine the ways in which, day in and day out, we accept illusory ideas about what's most valuable and necessary.

## NET INCOME

An important allure of "outside jobs" is the illusion of how much income is earned. That should be compared with what is actually "brought home." Let's say that someone is making $18 per hour at an office job. When you ask her how much she makes, she will tell you "eighteen dollars an hour." But take out taxes and deductions for medical plans, and the take-home pay is dramatically reduced. Then you need to factor in the other (often hidden, or taken for granted) expenses that job involves, including special clothes, supplies and equipment, commuting and travel, outside meals, and so on. Remember, too, you aren't working just eight hours for that paycheck. Many people in a big city spend huge amounts of unpaid time getting to and from their job. We know people who spend an hour and a half each way to commute—five (or six) days every week! Then there is the hour or so before you depart, getting washed and dressed. There are often the costs of gas, tolls, parking, and the many expenses of maintaining a car. And consider what else besides pay you take home. There are the many concealed costs to your health due to stress on the job, as well as the effects of working in a possibly unhealthy environment.

If you are completely honest with yourself about all of those auxiliary expenses, soon that $18 per hour doesn't add up the way you thought it would. Yet most people give away the best of themselves, all through their best years, for such a dubious return.

From what we've observed and read, it appears that the vast majority of regular eight-to-five workers hate their jobs but continue because they need the money. We can only assume that that terrible dislike for what one does with the majority of one's time during the cream of life is a major contributing factor to all sorts of sicknesses and diseases, including mental health, heart problems, stress, and possibly even cancer.

## SPENDING LESS

There are two basic ways to increase "buying power." You can earn more money, or you can spend less. Let's talk about how you can spend less.

*Coupons.* Though it took Christopher quite a while to come around to the idea of using manufacturers' coupons at the supermarket, Dolores has long been a "coupon cougar." More than fifteen years ago, Dolores began organizing her coupons into a three-ring binder designed to hold photos. Each page is a piece of cardboard covered on either side with clear plastic that can be lifted up to add or remove the coupons. She devotes a page to each category, for instance, salsas, or tofu, or oils, or coffee. From week to week on the appropriate pages she will also insert clippings from flyers or ads for any special offers. This way, when she goes shopping, she doesn't have a wad of coupons constantly falling all over the place as she thumbs through them trying to find the one she wants. She simply flips from page to page according to the store aisle and pulls out the coupons she needs.

Over the years, many other shoppers have noticed what Dolores was doing and initiated conversations. As time has gone by, we've noticed other people using the same or a similar system. Whether or not they got the idea from Dolores—we don't know.

Once, when Christopher was writing a weekly column for the Foothill Inter-City newspapers (a chain of four community papers), we both went coupon shopping for the purpose of writing an article on the experience. Dolores had saved all manner of coupons, and we went to a local supermarket one evening about ten o'clock. We deliberately went late, because we didn't want to feel "rushed," as it is easy to feel when you have coupons and people are waiting behind you in line.

It turned out that there was only one line open at the supermarket, and the man behind us was *very* impatient as we unloaded a cart of goods and began to produce our coupons. We wish we'd had a video camera to record his fidgeting, hand shuffling, and blatant eye rolling as his body language was saying, "Can't you speed it up?"

The total for the grocery bill was initially $54.11, but then Dolores handed the checkout line attendant a stack of coupons, all of which were double value. Dolores also had some "free" offers. The checker dutifully reduced the total bill, coupon by coupon, and we watched as the total got smaller and smaller. Even the man behind us became quite interested now that he saw that those coupons didn't mean just a few pennies off. He tried to get in as close as he could behind us to see what was going on, without seeming overly nosy.

When the grand total came to be $17.25, the checker said, "Wow! That's pretty good!"

This 68 percent savings was accomplished by using manufacturers' "cents-off" coupons, taking advantage of special offers, and shopping when a maximum number of the coupons were valid. We are also accustomed to watching for weekly specials at a particular store as well as making a point of shopping when a store has a policy of doubling coupon values and even accepting some triple-off coupon offers.

We once saw a woman on television who got coverage for purchasing more than $300 worth of groceries and paying about $7. That feat was the result of working with her "coupon club" (a form of coupon-sharing pool) and utilizing every possible rebate, discount, and special offer to get as many products as possible for free. What that woman did was by no means the norm, but her publicity stunt shows what *can* be done.

Dolores finds that she can *routinely* get 30 to 50 percent off the shopping bills by virtue of her discipline as a "coupon cougar." And she figures that she spends no more than ten to fifteen minutes a week clipping and filing coupons.

Dolores will not use a coupon or special offer to purchase an item that she would not normally use (or whose ingredients are not good), because that is false economy. But by taking the time to organize coupons by category into the three-ring binder, she can check the supermarkets' specials in advance against the coupons she has. Often she can buy a product on sale, then get a further reduction with a coupon. The binder system makes this quick and easy.

Also, manufacturers will periodically offer special deals, for example, "buy three, get one free." In those cases where the customer needs to save a proof of purchase, Dolores simply puts the label or receipt into the corresponding page of her binder until she has enough to cash in on the special offer. This avoids the problem of looking all over the house for wherever it was you stuck those receipts.

We used to be of the opinion that the only way to get big savings on food purchases was to buy in bulk, in huge containers or wholesale. Though we do on occasion buy groceries in these ways, we've found that buying in bulk is not always less expensive. You really need to compare the per unit or per weight prices; don't automatically assume that bigger quantities or case lots mean cheaper. Not infrequently we have noticed *higher* prices for items at those warehouse stores where everything is sold in big containers or by the case.

At other times, buying in bulk or buying cases not only saves money but saves you the extra time and effort of running back to the store to buy more of something. Plus, with crucial supplies and staple foodstuffs it's generally a wise practice to buy more than you need so as to have extra on hand in the event of an emergency. Amish and Mormon families are known for having several months' supplies on hand at all times.

Food storage is sometimes viewed as a selfish or "me-focused," excluding activity. The neighborly way to approach food storage is to follow the formula "six for six": each person in the household should have provisions for six people for six months. This means there's always plenty to share or barter. It doesn't mean that you should adopt the attitude that you're responsible to provide for others, but if you follow this method, you *could* do so if the circumstances warranted.

On occasion we have used coupons for products that we were stocking up on, and have paid about $10 for each $100 of retail value! To us, this seems like an integral part of economic survival. Both spiritually and economically, it is good to be frugal and thrifty—and good to allow society to support your efforts, which you can do by taking advantage of offers for discounts and free goods. As long as one does not become greedy, and doesn't cheat, there is no reason not to take advantage of all the offers and discounts that are easily and readily available.

### BARTERING

In addition to using coupons to get discounts and premiums, we have also found ways to trade for what we need or want. Trading, or bartering, is always somewhat complicated; the challenge is to accurately place a value on goods or services in order to get an even trade. Sometimes this kind of exchange works well, and sometimes not.

We have often traded goods or services for attendance at our classes, an arrangement that many participants have found acceptable and worthwhile. Our policy is that we never turn someone away due to lack of money. Yet if we just "give away" our talents, creations, and time with no recompense, our students are less likely to appreciate the value of what they are learning, and therefore tend to make less of a personal commitment to the class. We do have clearly defined fees for the sessions we teach, but we stay open to the possibility of having people do work here at our home (there is never a shortage of jobs to do) or trade crafts, garden produce, or something else with us.

One of the many ways we've benefited from working at bartering over time is that we've gradually realized that our needs are quite distinct from our wants. That is, usually the actual need is not "money," per se, but rather an object or a service. When we can equitably match the objects and/or services we offer with those offered by someone else, we all profit, and without having to resort to dealing with dollars. Which leads us to a another lesson learned. Although you cannot avoid money entirely in this society, dealing with others on a strictly monetary basis brings relationships down to the level of a mundane business transaction. Sometimes that's appropriate, but in many situations we find that there are "higher" avenues of interaction, where money plays no part. In a dollar-obsessed world, this may seem hard to explain, but those who have had the experience of volunteering for a good cause know that "work for hire" can't compare with the satisfactions of giving freely, by choice.

Some people constantly want to know "What did you pay for that?" as if the dollar expense defines the meaning of everything. It doesn't. By leaving money out of certain kinds of exchange, you thereby alter the interaction from a strictly mercantile one to a potential communing of different Selves.


Obviously, much of the time it isn't possible for us to function without cash, because most businesses aren't open to discussing trades or barters. You pay your power bill and water bill with a check. Period.

Still, we look for every opportunity to barter and often allow other people's discards to be our treasures. For instance, we rarely purchase firewood—we usually get that for free from people eager to dispose of prunings, trimmings, or logs from a tree they've cut down.

We buy no pharmaceuticals and no cosmetics. We do not buy or drink alcoholic beverages.

Though we are not certain how much money this saves, week after week we get salad greens, avocados, citrus fruit, and vegetables from our garden, as well as wild foods.

We watch for sales, discounts, and closeouts and always check the local listings in *The Recycler*, a Southern California newspaper filled with ads from private parties selling all sorts of things.

We occasionally stop at yard sales and have made some incredible purchases at yard and estate sales. We are not too proud to pick something up that somebody else threw away, nor are we above fixing, cleaning, and using it. We've found good tools this way, and also blankets, wood, fencing, furniture, and so on.


## TIME AND THE QUALITY OF LIFE

Many people today believe that they're spending all their time working, yet with very little in return. Unfortunately, such realizations may come too late to be remedied.

We think that the Amish people have the right idea when they keep their schools and work close to home. They don't have to go a long way to a job, thereby avoiding wasted time and energy, unnecessary expenses, and disconnection from their community. They can protect their families from undesirable influences, and there is the added bonus of having youngsters nearby where they can learn a trade from an early age. The Amish are firmly committed to valuing "quality of life" over all the stuff that modern society deems important or indispensable—car, home entertainment system, fancy clothes, foods bought for "convenience" and prestige rather than fresh garden flavor and nutritional value.

Christopher remembers asking an Amish craftsman in Ohio how he was able to "get by" when asking such low prices for his products: clocks, furniture, children's toys, and more. He was thoughtful for a moment, then pointed out that he only wanted a "fair price" for his wood products, not an outrageous sum. He then smiled broadly and commented that most urban "Yankees" (a term for any non-Amish person) probably earn more money at their jobs, but then they have to buy everything at high prices, including work clothes, transportation, and so on. "And because they rarely exercise," he added, "they even have to pay to go somewhere to do that." He thought this last point was hilarious. "I don't use electricity," he then said with a stern face, "and I don't avoid labor." His work was his exercise, and because his way of life enabled him to achieve more than one end with each activity, he was effectively getting more for less. This was his way of telling Christopher how he was able to "get by" for the prices he charged.

When one thinks of homesteading in remote rural areas or in the wilderness, it's inspiring to imagine such homesteaders providing for many or even most of their needs with what they can grow, make, or find in nature. As we've learned, to a much greater extent than most people recognize, the same can be done in an urban setting as well, but you have to work harder to achieve self-sufficiency, and there are very few examples to emulate. Urban homesteaders are therefore pioneers in the urban wilderness.

*Christopher (with vest) leads a wild food outing in the Arroyo Secco.*

During an early morning discussion with our friend Vernon, we brought up the pros and cons of having a regular eight-to-five job. Christopher shared that during the time of his highest-paid "full-time" job, he always felt as if he were dying inside, spiritually. Sure, there was the so-called security of getting money on a predictable basis, but that job became nearly the totality of his life.

Vernon has often attempted to help us see other possibilities, ways to earn our needed income by doing what we love, by sharing what we believe in, and by marketing from home the various products and services we can offer. It might sound as if he was urging us in the direction of barter, but that word doesn't begin to explain the richness that becomes available to those who choose another way of living, another way of looking at the world.

He remarked that the musical group that he played with would begin and end each performance with a lively rendition of "Love Makes the World Go Round." Yet he always felt that this was very hypocritical, because it isn't "love" that make *this* world go around, but rather fear. He explained to us that fear drives most people for decades to hold down regular jobs that they hate, because they "need the money." This, of course, opened up a whole can of worms—and led us into long discussions about the differences between "needs" and "wants," "cost" and "real value," and about how it is rarely money, per se, that we need.

Vernon's view was that by letting fear control our lives, we are eternally cut off from the real magic of the world. For example, he shared a story of a friend who had found two hard-to-obtain books in a local trashcan. She had actually been looking for those books, as they contained vital information she needed in her personal life. She had been complaining about not having enough money to buy the books or other necessities of life.

Vernon told us that such "accidents" contain within themselves an expressive language, if we open ourselves up to reading that language. In this case, the woman's accidental find came at a time when the "I'm poor in dollars" way of thinking was luring her to the brink of despair. Yet here came this unique coincidence, the books discovered in the trash being gifts that were part of her reward for good faith and conscientious efforts in her life. She didn't find a wad of money in the trashcan, but something ultimately much better: as she studied those books, they provided clues allowing her to take the first steps leading her out of the dollar doldrums.

Vernon then went on to tell us a story about his own family. Many years earlier, Vernon's wife had felt compelled to give up a job in order to return to and finish college. Vernon was self-employed, and his income was very sporadic. Yet by keeping open to opportunities, they had found themselves living in a Greene & Greene–designed mansion on a 3-acre estate, with two pedigreed dogs and a sparkling 1947 Ford. "This combination gave us the appearance of being wealthy eccentrics," he told us.

Around that time, Vernon noted an interesting phenomenon. In order to have a greater monthly supply of "disposable" dollars, one could choose either to *earn more* and/or *spend less*. "I know the idea sounds too simple to give it much thought, but just think *into* that choice awhile!" he urged us. "I decided to make a science of the latter alternative— spending less."

For example, he had begun to think of television as an important medium for learning, if people took the time to use it correctly. The 14-inch TV that had been fine for the small apartment they'd just moved from was now inadequate for their giant new living room. What was the logical recourse, considering that Vernon was self-employed with irregular income and his wife was going back to school? They didn't have the cash to just go and buy a bigger television. Should they make do with the too small but still working one they had? Should they complain and feel bad? Should they do as most folks do, and go into debt for a big new one from Sears?

For several years, Vernon had been finding unbelievable bargains on supermarket and laundromat bulletin boards. He began focusing on ads for televisions, and before long he saw one for an almost new color TV with a remote control for "$750 or best offer." Vernon asked us, "If you were in my exact circumstances, what would you have done?"

He said that his wife was immediately discouraged. "It's impossible! Just forget it." But then there came into play that magic of which this world is really composed, for those willing to be open. Vernon began to think of ways he could "lead with an offer." He made a list all the items they owned but didn't need. Then, with that list in hand, he called the phone number on the ad for the TV and made personal contact with the seller. He turned out to be an Australian man who was doing very well as a self-employed gardener, and who preferred living in a tiny guest cottage on one of his customer's estates.

During their conversation, it came up that he needed an economical vehicle for hauling his lawn mower and other tools. It just so happened that Vernon and his wife had an Austin A-40 station wagon, which got 52 miles per gallon—proved on a recent trip—and which they didn't really need to keep. As an Australian, this man well knew the virtues of the Austin A-40, a popular vehicle in Commonwealth countries. The man acknowledged to Vernon that the TV needed to "warm up" before a picture clicked in. Vernon acknowledged that the Austin used a bit more oil than it should. Even so, the man eagerly sought to consummate the transaction.

As they signed over their respective chattel, they gave one another verbal guarantees of the good working condition of the items traded, and they continued to exchange valuable information that each would need in order to make best use of his new possession. The Australian even offered to come over and adjust the TV in Vernon's living room, and Vernon and his wife invited him to stay so they could all have dinner together. His bachelor's countenance beamed at that proposition! While he was at their house, he identified a number of valuable plants around the yard, and he showed Vernon how to do some pruning. Everyone ended up very happy, and they even phoned each other regularly over next six to eight months to check on the other's satisfaction. How many regular merchants would do that?

"Of course," Vernon told us, "in place of all the initial 'wasted time' we spent making such contacts, my wife and I could have just gone out and found high-paying jobs in order to purchase a TV on a time plan from an impersonal salesclerk. But by using this same technique, over the next two years we acquired a Norton motorcycle, a 1957 Chevy panel truck, an excellent bedroom suite, and a full set of tools, along with many other items."

An essential part of Vernon's "science" was that even during a time of irregular income, Vernon and his wife kept up their support of worthy charities (even if only donating $1 dollar); continued to put money into savings, and made effective investments in the stock market. They also made time to research income tax and nonprofit incorporation laws, the intricacies of bank and credit unions, international finance, real estate, and so on.

Here is another example of making the most of accidents or coincidences. Vernon had seen an ad in the paper for a proofreader's job at the paltry rate of $2.50 per hour (remember, this was thirty years ago). "In such a situation," Vernon pointedly asked us, "what would each of you have done? I felt the job was ideal for me, so I applied and was hired. After I was in the job five or six months, I was given added responsibility for reading classified ads. . . . . Note the sequence: I had to *first* prove my abilities and dedication with 'no promises' before I was shown the next-level opening."

In the classified department, Vernon read listings describing for-sale items forty-eight hours before the paper came out on the street. This opened a new world of possibilities. Using this new means of searching for deals, Vernon and his wife purchased a Mercedes 220S, a Corvair pickup, the property where they now reside, and two other properties. "Of course," Vernon admitted, "I could have gotten a proofreading job for at least six dollars an hour in Pasadena or in downtown Los Angeles." However, he felt that valuing the hourly wage more highly than the range of benefits offered by the lower-paying (and closer-to-home) job would have represented a missed opportunity—and a spiritual side-track.

What did we learn from Vernon's good example? Allowing a "poor, destitute me" self-image to overwhelm imagination and creativity is like letting a dam hold back a lake—you could actually die of thirst when meanwhile there is a huge reservoir of water only fifty feet away.

Our objective conclusion is that most of the industrialized world's adults have chosen to "die of thirst" so they can have the security of a "good job," "good income," and "good retirement plan." And because Mammon does dribble out such rewards, it is possible to live one's entire life there in that allegorical desert and never even be slightly aware of the awesomely vast waters so *close* at hand! This is the tragedy of modern human existence.

Once, during a period of homelessness before we were married, Christopher was engulfed in thoughts of "poor me" and "I'm destitute," and he could scarcely see a way out of the darkness. Dolores provided him with a simple set of practical tools anyone can use if only they choose to do so. Here are four "magic" ways to improve your financial situation:

1. Never waste anything.
2. Continually improve your personal honesty.
3. Leave every situation or circumstance better than you found it.
4. Tithe to the church (or organization) of your choice.

We know that these are genuine, practical solutions. We have heard people say that they cannot make these efforts—such as tithing, or improving an environment—because "We are poor." Our perspective is that they have their reasoning backward. They are poor *because* they do not engage themselves in the world in these ways. Logical thinking leads to erroneous conclusions when the premise is false.

It was around 1984. Ronald Reagan was U.S. president, and Mikhail Gorbachev was in charge of the Soviet Union—the Iron Curtain had not yet parted. And although not everyone still remembers, there were at least two occasions during that period when the saber rattling between Reagan and Gorbachev was deemed by military experts to be truly serious, so serious that we were indeed faced with the potential of a nuclear showdown.

It was in this global context that Gilbert Nyerges, Christopher's oldest brother, a longtime commodities investor, became concerned for his family. He went out one weekend, after he had heard various rumors and rumblings, and purchased a hundred cases of Mountain House camping foods, which cost a small fortune. His garage looked like a warehouse. He soon added bottled water and water filters to the stockpile.

The menacing disagreements between the superpowers lessened, and the two leaders actually became friends, opening a new era in modern history. The Cold War was over, declared the pundits. The bombs never fell. And Gilbert Nyerges had not opened a single box of the survive-the-nuclear-war food cache in his garage.

That is, he didn't open a single box until 1993. At that time, the financial markets were in a slump, real estate values were in a deeper slump, and Gilbert's income had been reduced by nearly 80 percent. Then there was his divorce, necessitating payments to his ex-wife and to a lawyer. This was followed by the tragic death of his youngest son due to drowning. Realizing that his once expensive home was now worth approximately $150,000 less than he had paid for it—and for which he was still paying, with a sizable monthly mortgage—Gilbert and his seven children moved to a smaller home.

With the family income so drastically fallen, Gilbert recognized that at least one meaningful contribution to the solution was sitting there in his garage. It was not a war or earthquake or nuclear attack that finally got him to begin using the large supply of stored food, but personal hard times. "And it really wasn't that bad," he said with a grin. "In fact, most of it was actually pretty good."

"We began with the powdered milk and fruit juices," he explained, "using the milk in cereal and the juices for breakfast. And we had plenty of canned granola. We also had dried eggs and cheddar cheese, and the children loved that. Then we began opening the dried veggies, which we'd use as the vegetable course for our dinners, first soaking them in water, then warming them up. They weren't exactly like fresh vegetables but they really were delicious. Even the children ate them." He smiled, remembering how they served beans, peas, corn, broccoli, mashed potatoes, diced carrots, and other reconstituted vegetables with their meals. The storage was loaded with foods they actually enjoyed.

Gilbert would also take dried fruits and mix them in sandwich bags for the children's lunches—peaches, apples, pears, nectarines, strawberries, and other varieties. They would also eat the dried fruit as snacks or soak them in water to use as dessert with the evening meal. They particularly liked the dehydrated fruit cocktail.

Gilbert has come to appreciate the wisdom of storing provisions—just as the Amish and the Mormons do—toward the possibility of some unforeseen emergency. And that emergency need not be anything as severe as a war, or weather catastrophe, or "the end of the world as we know it." Unemployment or a downturn in one's financial prospects can also be devastating, especially if you have children who depend on you.

Although Gilbert confides that he still doesn't know what an MRE is (ready-to-eat military meals), he has become aware of some wild food plants and has taken steps to teach the children how to begin adding that valuable resource to their diet. He feels fortunate that he had the foresight to purchase dried food when he had the opportunity and surplus income. He now knows much more about what it feels like to live off one's own resources, as you'd have to do in the aftermath of a major earthquake or other natural or human-caused catastrophe.

Eight years after their dramatic change in fortunes, Gilbert and his children still have about twenty cases of the emergency food supplies. He says that their stockpile of dried food really tided them over through their difficulties and that all of it was quite good once they got used to the differences in flavor.

As Alexander Graham Bell said, "When one door closes, another opens, but we often look so long and regretfully upon the closed door, we do not see the ones which open for us."

# E p i l o g u e

We hope you've enjoyed this sharing of how anyone can live more ecologically on a limited budget, even in the city.

We think we've succeeded in accurately depicting our efforts at urban homesteading. We don't claim to have all the solutions to challenges that will arise. And, of course, we have had many failures and problems as we have attempted to "live lightly on the earth." We don't want the reader to think that this was a simple or an easy path. But generally our failures and obstacles caused us to rethink and revise our actions, or, in some cases, to stop doing something altogether. As a wise man once said, "The road to success is paved with the cobblestones of our failures." We can vouch that we walk that cobbled path.

We also want to mention that, although in this book we have discussed a lot of what is involved in living ecologically in the city, we did not have room to cover every strategy. For example, though we mentioned how we earn a living, there are many ideas and possibilities for home businesses that we have not introduced at all. Your options will depend on your individual circumstances.

In addition, we haven't discussed issues such as home security, alarms (electronic and primitive), retrofitting your home to deal with disasters (earthquakes, tornadoes, floods, and even manmade disasters such as riots), firearms ownership and protection, the value of neighborhood co-ops and meetings, Neighborhood Watch groups, alternative communication systems such as ham radios, and so forth. In other words, there is a whole range of other topics that you must thoroughly investigate if you intend to live well in the city and be a vibrant part of your community.

Finally, we are strong advocates of finding a way to know your neighbors and be a positive part of your neighborhood—just as we assume most people did generations ago. People say they no longer practice "old-fashioned" neighborliness out of fear, the idea of not enough time, and the weak excuse of "my little efforts won't make a difference in the big city."

We believe everyone can be a part of a larger solution if we only choose to do so. Though everyone believes that they "don't have enough time," we feel it is imperative to make the time to do what you feel is important in life.

Our urban homestead lifestyle is part of our on going day-to-day life. We welcome your questions and comments.

# Resources

EDUCATION

School of Self-Reliance, Box 41834, Eagle Rock, CA 90041, www.self-reliance.net, is the organization founded by Christopher and Dolores Lynn Nyerges. They offer classes throughout the year on topics ranging from wild food identification, outdoor survival, solar oven construction, and home energy self-reliance. The schedule is listed on Dolores's Web site. They also publish a newsletter and many other booklets and books. Their publication list is available by writing to them, or on-line.

WTI, Inc., is a nonprofit educational and community service organization, incorporated in 1971. They offer a broad diversity of programs, including Holy Day and holiday gatherings, spiritual counseling, numerous classes, and ongoing experiments and testing programs in which volunteers may participate. Contact WTI via School of Self-Reliance, Box 41834, Eagle Rock, CA 90041.

BOOKS

Rodale Press (Emmaus, PA) is another publisher of magazines and books for like-minded folks who are reading this book. They publish *Bicycling, Prevention,* and *Organic Gardening* magazines, and a complete line of gardening and homesteading books.

PRODUCTS

Real Goods is perhaps the best single source for alternate energy supplies, and information. Contact them at the Solar Living Center, P.O. Box 836, Hopland, CA 95449. For a free catalog, call 1-800-762-7325, or visit their Web site, www.realgoods.com.

*Home Power* magazine is another excellent source for the hard-to-get information needed by urban homesteaders who wish to generate at least some of their own power. Available at newsstands and on-line.

Iodine crystal water purification kits, mentioned in chapter 6, are available from School of Self-Reliance, Box 41834, Eagle Rock, CA 90041. Also check their Web site at www.self-reliance.net.

# Bibliography

Aebi, Ormond and Harry. *The Art and Adventure of Beekeeping* vol. 1. Santa Cruz, Calif.: Unity Press, 1979.

———. *Mastering the Art of Beekeeping* vol. 2. Santa Cruz, Calif.: Unity Press, 1979.

Auerbach, Les. *A Homesite Power Unit: Methane Generator.* Madison, Conn.: Alternative Energy Systems, 1979.

Belanger, Jerome. *The Homesteader's Handbook to Raising Small Livestock.* Emmaus, Pa.: Rodale Press, 1974.

Dastur, J. F., F.N.I. *Medicinal Plants of India and Pakistan.* Bombay, India: D.B. Taraporevala Sons & Co. Ltd., 1988.

Davidson, Joel, and Richard Komp. *The Solar Electric Home: A Photovoltaic How-to Handbook.* Ann Arbor, Mich.: Aatec Publications, 1983.

Fukuoka, Masanobu. *One Straw Revolution.* Emmaus, Pa.: Rodale Press, 1978.

Graham, Joe. *The Hive and the Honey Bee.* Hamilton, Ill.: Dadant & Sons, 1975.

Halacy, Beth and Daniel Stephen. *Cooking with the Sun.* La Fayette, Calif.: Morning Sun Press, 1992.

Halacy, Daniel Stephen. *Fun with the Sun.* New York: MacMillan, 1959.

Jarvis, D. C., M.D. *Folk Medicine: A Vermont Doctor's Guide to Good Health* New York: Henry Holt, 1958.

Jenkins, David and Frank Pearson. *Feasibility of Rain Water Collection Systems in California.* Davis, Calif.: California Water Resources Center at the University of California, Davis, n.d.

Komp, Richard J. *Practical Photovoltaics: Electricity from Solar Cells.* Ann Arbor, Mich.: Aatec Publications, 1984.

Langer, Richard. *Grow It!* New York: Avon Books, 1974.

Murchie, Guy. *The Seven Mysteries of Life: An Exploration in Science and Philosophy.* Boston: Houghton Mifflin Company, 1978.

Potts, Michael. *The New Independent Home.* White River Junction, Vt.: Chelsea Green, 1999.

Rodale, J. I. *The Complete Book of Composting.* Emmaus, Pa.: Rodale Press, 1960.

*Rodale's All-New Encyclopedia of Organic Gardening,* Fern Marshall Bradley and Barbara W. Ellis, eds. Emmaus, Pa.: Rodale Press, 2000.

Roy, Rob. *The Sauna.* White River Junction, Vt.: Chelsea Green, 1996.

Shaeffer, John, ed. *Real Goods Solar Living Sourcebook,* 11th ed. Hopland, Calif.: Gaiam Real Goods, 2001.

Schaffer, P. S., William E. Scott, and Thomas D. Fontaine, "Antibiotics That Come from Plants." *Yearbook of Agriculture,* 1950–51, 732.

Shealy, C. Norman, M.D. *The Illustrated Encyclopedia of Natural Remedies.* Dorset, U.K.: Element Books, 1998.

*Sunset Western Garden Book,* Kathleen Norris Benzel, ed. Menlo Park, Calif.: Sunset Books, Inc., 2001.

Terre Vivante. *Keeping Food Fresh: Old World Techniques and Recipes.* White River Junction, Vt.: Chelsea Green, 1999.

Walters, Charles Jr. and C. J. Fenzar. *An Acres USA Primer.* Raytown, Mo.: Acres USA, 1979.

Wilson, Alex, Jennifer Thorne, and John Morrill. *Consumer Guide to Home Energy Savings,* 7th ed. Washington, D.C.: American Council for an Energy-Efficient Economy, 1999.

# Index

garlic, 89, 93, 95
gas, natural, 2, 134, 135, 140
geese, 25, 26, 28, 36, 59–60
glass, recycling, 175–79
glass containers, uses of, 110–11
gloves, 191
grapefruit and grapefruit trees, 14
grapes, 13
grass clippings, 78, 87. *See also* mulches
graywater, 112, 113, 114, 119, 190
greens, wild, 32–35
grief and mourning, 68–70

Hafenfeld, Mel, 70
hair dryers, 128
Halacy, Dan and Beth, 144
Hall, Ellen, 180, 183, 205
Hall, Timothy, 43, 177, 189
hand tools, 126
health, 94–98. *See also* medicinal plants
heating systems, 125, 127, 129–30, 133–49.
    *See also* water heaters
herbicides. *See* chemicals
herbs, 9, 10, 89, 95
"hobo" lanterns, 183
"hobo" solar water heaters, 181–82
Hogeboom, Geraldine and Ernest, 82, 87,
    139, 195, 201
home offices, 171–72
homesteading. *See* urban homesteading
honey, 43, 48, 53, 56–59
house design, 1–5
Hoyer, Robert, 54
human feces, 86, 1221
hummingbirds, 42–43, 90

iceboxes, 150
ice–cream cartons, 197
incandescent lightbulbs, 129
income, sources of, 210–14, 217–18
insects, 44, 82, 87. *See also* pests and pest
    control
insulation, 127
integral gardening. *See* gardening
"intertie" solar electric power, 165
inverters, 165, 166, 169, 172
iodine, water and, 102
"iron water," 97, 180

Jarvis, Dr. D.C., 57
Jenkins, David, 107
Jenkins, Joe, 86
Johnson, Edson, 188
Jonke, Thomas, 85

jugs, plastic, 190

Kallman, Stefan, 104
kapok, 19, 86
killer bees, 55
kitchens, 140–48
kitchen scraps. *See* garbage
knife sharpeners, 126, 128
Komp, Richard, 165

lacewing flies, 92
ladybugs, 91, 92
lamb's-quarters, 12, 32, 33, 42
land ownership, meaning of, 96
Langer, Richard, 45
lanterns, 183
lavender, 9
lawn mowers, 128
lead, in water, 108
leaf blowers, 128
lemons and lemon trees, 20
lights and lighting, 126, 127, 129, 167
lime and liming, 82, 84
"logrollers," 199–200
luminarias, 198

Madeira vine, 12
mail, recycling, 194–95
malathion, 91–92
mallow (cheeseweed), 30, 36, 40
manures, 24, 45, 84–86. *See also* worms and
    worm castings
"manure tea," 84–85
marshmallow, 38
McCorison, Mike, 101
medicinal plants, 24–25, 26, 38. *See also*
    honey
Mediterranean fruit flies (medflies), 91
mesh bags, 189
mess kits, 180, 182
metals, recycling of, 97, 179–87
meters, electric, 165, 172
methane, 121, 163
microwave ovens, 128, 140
milk cartons, 196
mint, 12
mixed paper, recycling of, 194
molasses, 209, 214–17
monoculture farming, 71, 89, 93
Mormons, 221, 230
mulches, 12, 79, 84, 87–88, 199, 201, 204
Murchie, Guy, 51, 99
mushrooms, 88
mustard plants, 32